Harvard
Business
School

Executive
Education

Smart Rivals

How Innovative Companies
Play Games That
Tech Giants Can't Win

Smart Rivals

FENG ZHU
BONNIE YINING CAO

HARVARD BUSINESS REVIEW PRESS
BOSTON, MASSACHUSETTS

Library of Congress Cataloging-in-Publication Data

Names: Zhu, Feng (College teacher), author. | Cao, Bonnie, author.
Title: Smart rivals : how innovative companies play games that tech giants can't win / Feng Zhu and Bonnie Cao.
Description: Boston, Massachusetts : Harvard Business Review Press, [2024] | Includes index.
Identifiers: LCCN 2023057182 (print) | LCCN 2023057183 (ebook) | ISBN 9781647826048 (hardcover) | ISBN 9781647826055 (epub)
Subjects: LCSH: New products. | Strategic planning. | Technological innovations.
Classification: LCC HF5415.153 .Z58 2024 (print) | LCC HF5415.153 (ebook) | DDC 658.5/75—dc23/eng/20240402
LC record available at https://lccn.loc.gov/2023057182
LC ebook record available at https://lccn.loc.gov/2023057183

ISBN: 978-1-64782-604-8
eISBN: 978-1-64782-605-5

From Feng
To Mom, Dad, Ping, Evan, Xiaomei, and Dazhuang
From Bonnie
To Mom, Dad, Charles, and Eureka

CONTENTS

Smart Rivals

Fight the Right Battle

How can traditional businesses survive and thrive in their battle against tech giants? If you've picked up this book, you're as intrigued by this question as we are. Over the past decade, traditional businesses across various sectors around the world have experienced continuous disruption and encroachment from digital startups and tech giants.

Traditional financial institutions are grappling with the rising fintech companies, which are reshaping every aspect of business, from payment services to consumer lending. In the automotive industry, traditional car manufacturers are competing with innovative tech players, such as Tesla, and autonomous vehicle pioneers like Google. The retail sector is under threat as tech behemoths, from Amazon in the United States and Alibaba in China, expand into the physical retail space, posing a direct challenge to traditional businesses. The advent of large language models such as OpenAI's ChatGPT, Google's Gemini, and Meta's LLaMA has generated both enthusiasm and concern among traditional businesses. While these technologies are exciting, they may further strengthen the power of tech giants.

Recognizing the power of digital technologies, in the past decade anxious executives have come to Harvard Business School (HBS) to learn how to digitally transform their businesses. We share with them what has driven tech giants' success and urge them to embark on digital transformation within their own organizations.

Not surprisingly, many of these executives have become fluent in the strategies of tech giants and have actively adopted digital technologies. They've developed mobile apps, implemented omnichannel strategies to reach users both online and offline, hired data scientists to gather and analyze data, and begun running experiments like those of Meta and Amazon.

The more resourceful ones have gone a step further, investing in or incubating their own startups to compete directly with tech giants. Yet, despite substantial efforts and investments in digital transformation, many still struggle to prosper amid competition from digital titans.

Why is it so difficult for traditional businesses to flourish in the digital age? Our research suggests that a key problem is that many traditional businesses are *fighting the wrong battle*. They envision that the future's triumphant product or service within their industry will combine digital and traditional elements. As tech giants grasp these traditional elements to encroach upon their industries, they must outpace them by mastering digital

FIGURE I-1

3

Introduction

The race to win the battle

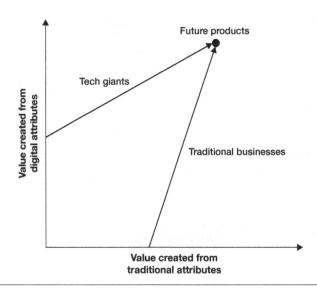

capabilities before these giants can attain proficiency in traditional aspects, thus securing a competitive edge (figure I-1).

For instance, a grocery chain may find itself in a competition to harness data and digital channels with greater efficiency than Amazon can master the operation of physical grocery stores.

This mindset is misguided. Drawing on more than a decade of research and case studies conducted on companies across the globe, we've discovered that traditional businesses thriving in the digital age often play smart against their digital adversaries. Instead of trying to outpace tech giants, they craft their own path and engage in a game that tech giants simply can't win (figure I-2). We call these innovative traditional businesses *smart rivals*.

FIGURE I-2

Crafting a new path

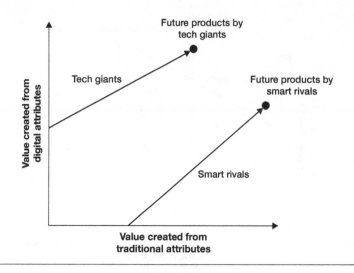

These smart rivals employ digital technologies to amplify their existing competitive advantages. They drive consumer centricity and often build platforms and ecosystems in ways that are uniquely their own. In the end, they develop products or services that are profoundly different from those offered by tech giants, consequently making them difficult for tech giants and other competitors to emulate.

Smart Rivals versus Superpowers

This book is crafted to guide companies of any size or age as they strategize their approach, aiming to become smart rivals in a world dominated by tech superpowers. Undeniably, traditional businesses still need to embrace digital transformation. However, prosperity in the digital age extends beyond mere transformation—digital transformation should be viewed as a toolkit, not the ultimate objective.

The objective should be to build competitive advantages by delivering unique product features and benefits that tech giants and other competitors cannot replicate. It's imperative for traditional businesses to study tech giants. Yet the intention of this learning is not to imitate, but to understand how they can distinguish themselves and harness the infrastructures established by these tech behemoths to amplify their own competitive edges.

By doing so, traditional companies can grow without engaging in a zero-sum race against tech giants, a race that countless traditional businesses deem unwinnable. Forging a new path requires the right direction and a touch of creativity. This is uncharted territory for traditional businesses, with insights not readily available from data analytics or by observing tech giants. Yet, this path is eminently feasible, based on our research studies and consulting work with traditional businesses globally.

The Path Ahead

Unlike many management books that focus on tech giants, our work showcases creative paths that various traditional companies across industries have traveled. These organizations, many founded before the advent of the internet, span a broad spectrum. Examples include Domino's, a major US pizza chain; Powell's Books, the world's largest independent bookstore; and Sephora, the French multinational beauty product retailer.

At our home institution, HBS, we emphasize real-world business challenges and decisions in our teaching, utilizing the century-old case method. In this book, we explore what we call smart rival strategies through case studies, including many companies we've personally visited and whose executives we've interviewed.

More than half of the examples in the book feature companies based outside the United States. These include Belle, the leading women's footwear retailer in China; DBS Bank, Singapore's premier consumer bank; EbonyLife, Nigeria's top media conglomerate; Telepass, Italy's popular electronic toll payment service; and Zé Delivery, a beer delivery platform incubated by Ambev, Brazil's renowned brewing company.

Forging a new path calls for leaders to follow certain principles that continually challenge their conventional thinking, driving them to become innovative competitors instead of simply followers of tech giants. This book offers overarching principles and strategies that will empower you to establish yourself as a smart rival across six key areas.

In chapter 1, we invite you to look inside your business to understand your company's strengths rather than obsessing over shifts in the external environment. This chapter explores how companies can leverage digital technologies to amplify existing competitive advantages, deepening their unique position and rendering tech giants unable to win the competition. It emphasizes the importance of determining the right innovation direction in the digital age. Without a clear innovation path, your strategies risk becoming aimless amid the flurry of new technologies frequently introduced by tech giants. Failing to set a clear path of innovation could leave you in a perpetual state of playing catch-up, rapidly diminishing your competitive strength.

Chapter 2 focuses on customer centricity. Tech giants have earned their reputation for their exceptional focus on customers, leveraging their immense data resources to deliver highly personalized offerings. In this chapter, we illustrate how traditional businesses can creatively leverage their inherent traits to overcome data disadvantages and adopt customer centricity in their unique ways.

Chapter 3 explores how traditional businesses can uncover opportunities to create platforms within their own organizations. Platform business models aren't exclusive to tech giants. By building innovative platforms from their existing resources, traditional businesses can harness network effects and build competitive moats in their ongoing rivalry against tech giants.

Moving beyond platforms, chapter 4 explores how smart rivals can establish and rapidly scale their own ecosystems to rival those of the tech giants. We examine how these companies can involve tech giants as participants within their ecosystems to foster growth. Our analysis reveals a key distinction: unlike tech giants, traditional businesses need not occupy a central hub position within their ecosystems to flourish and prosper. By strategically orchestrating their ecosystem dynamics, traditional businesses can harness their unique strengths and leverage the collective power of diverse participants to drive innovation, competitiveness, and sustainable success.

Chapter 5 delves into the intricate relationships between traditional companies and tech giants. Today, many traditional businesses find themselves engaged in both collaboration and competition. For example, UPS aids Amazon in package delivery while also competing with Amazon's delivery service. Similarly, Best Buy formed a partnership with Amazon in 2018 despite being competitors. This chapter pinpoints potential conflicts in such relationships and discusses strategies that smart rivals can use to strategically mitigate risks when partnering with tech giants. These strategies encompass a spectrum of approaches, ranging from strategic channel bridging to leveraging regulators.

Chapter 6 confronts the reality of disruption. Traditional businesses are continuously at risk of disruption by emerging or established tech giants. While the strategic moves discussed in earlier chapters aim to minimize this risk, the possibility of disruption persists due to ongoing technological advancements and other shifts in the external environment. In this chapter, we outline a set of general principles that traditional businesses can employ to bounce back and successfully revitalize their enterprises in the aftermath of disruption.

Finally, the conclusion provides a list of questions designed to help traditional businesses begin their transformation into smart rivals.

. . .

Digital technologies are often viewed as threats, but they are a blessing in disguise for smart rivals. They bring opportunities that can unlock the potential of traditional businesses, catapulting them to heights previously unattainable. We invite you to join us on this transformative journey. Let's evolve into smart rivals, together.

CHAPTER 1

Amplify Your Strengths

Wind extinguishes a candle and energizes fire.

—Nassim Taleb

Burgeoning digital technologies coupled with the Covid-19 pandemic have drastically reshaped our eating habits. At the start of the twenty-first century, options for restaurant food delivery were largely restricted to staple choices like pizza and Chinese food. However, the rise of digital food-delivery platforms such as DoorDash, Grubhub, and Uber Eats has broadened our restaurant choices and offered a convenient method for ordering food. These platforms, offering enticing promotions and subsidies to consumers, surged in popularity as restaurants were forced to shutter during Covid-19 lockdowns.

Yet, the fierce competition and the increasing popularity of these platforms did not seem to hinder the growth of one pizza restaurant chain—Domino's. This chain outshone even the world's largest tech companies, with its shares closing out 2010–2023 with a total return of over 4,600 percent, surpassing Amazon, Apple, and Alphabet (Google's parent company). So, what is Domino's growth recipe that has outpaced these tech giants?

Domino's remarkable growth is especially noteworthy given the struggles many traditional businesses face against tech giants. The demise of companies like Bed Bath & Beyond and Neiman Marcus, among others, has highlighted traditional businesses' vulnerability in the uphill battle against digital superpowers. The Covid-19 pandemic further intensified the competition on all fronts as more customers transitioned from offline to online purchases, allowing tech giants to capitalize on the struggles of non-digital natives.

As tech giants continue to exert greater influence across various industries, the question for traditional businesses is no longer whether to respond. The critical and immediate question at hand is: How do you best respond? Incumbents typically try to counter these powerhouses by emulating them, either through self-transformation or by establishing new business units. However, this mimicry approach, aimed at defeating tech giants at their own game, often falls short. Among various reasons, it's simply unrealistic to expect traditional businesses to develop technical capabilities on par with tech giants, such as devising superior AI algorithms or building advanced tech infrastructure to support extensive user bases and data demands.

In addition, tech giants typically operate on business models that diverge significantly from those of traditional businesses in terms of creating and capturing value. Shifting to such business models demands major organizational change, a feat most organizations are not capable of. More critically, the more a firm's strategies resemble another's, the fiercer the competition becomes. In such head-to-head confrontations, few traditional firms have the financial resources of tech giants, who can afford to endure extended periods of high expenditure.

It is noteworthy that even tech superpowers themselves need to differentiate to thrive. For instance, Microsoft's Bing historically struggled to compete with Google due to a design too similar to Google's. Bing's competitive position only significantly shifted after it integrated OpenAI's large language model, offering a more differentiated search experience.

Our research suggests that smart rivals often embrace a different strategy against tech giants. Rather than attempting to defeat tech giants at their own game—a strategy that will inevitably lead to an endless chase to match the superior technical prowess of tech giants, consequently diminishing their own competitive advantage, smart rivals forge their own growth paths based on existing strengths.

Amplify Your Strengths

We take an in-depth examination of two traditional businesses, Domino's and Sephora, in this section. As they illustrate, smart rivals amplify their existing strengths to a level that tech giants struggle to compete with. Many thriving traditional businesses in the digital age share this common trait.

Domino's: Differentiating through order fulfillment

The growth of various food delivery platforms that offer a wide range of cuisine, including pizza, has posed a significant threat to Domino's business model. So, how did Domino's address this threat and stay competitive?

In today's dynamic food industry, establishments like Domino's have numerous strategies to consider. Beyond traditional dining-in, drive-through, and pick-up services, these restaurants can collaborate with third-party platforms or outsource to white-label delivery service providers like Olo and Relay—third-party providers that work with restaurants and appear as if the service is handled in-house. As food delivery surged during the pandemic, restaurants could also transition to ghost or dark kitchen models, eliminating physical restaurant locations and relying solely on delivery to serve their customers. Amid the hype of the metaverse, restaurants could open virtual outlets that offer home delivery: customers could place orders from a virtual restaurant and have it delivered to their doorstep in the real world. Given this vast array of options, what strategies have set Domino's apart from peer restaurants when responding to these tech giants?

Businesses across various industries grapple with similar dilemmas today. To chart a path forward, it is helpful to evaluate the fundamental value your business provides for its customers. What aspects of your business do your customers value most? In many industries, consumers' purchasing decisions are affected by a limited number of key factors. In the restaurant business, diners' decisions largely depend on a handful of simple factors: variety, price, taste, and order fulfillment experience.

As a pizza restaurant chain, Domino's is unlikely to offer the same breadth of choices as food delivery platforms, which aggregate offers from multiple restaurants. Pricewise, Domino's also finds it challenging to compete with delivery platforms that offer substantial subsidies to customers to spur growth (at least until they establish some market dominance). Regarding taste, Domino's, despite its popularity, hasn't always had the best reputation for pizza quality. In fact, prior to 2008, Domino's had long relied on the same recipe and was compelled to alter it after recurring customer complaints about their pizza crust's cardboard-like taste.[1]

Historically, Domino's key competitive edge was its expertise in order fulfillment. It has been known for doing its own delivery for over fifty years. In the company's early days, delivery drivers used Domino's cars emblazoned with the magnetic logo decals, while today more drivers use their own vehicles. In 1984, Domino's advertised the "30-minute delivery, or it's free" guarantee, which drove Domino's rapid growth. By the 1980s, this once-small chain had grown to encompass five thousand stores before the decade ended.

However, Domino's journey became increasingly tumultuous before the digital era. In 1993, following multiple settlements from million-dollar lawsuits stemming from car accidents involving its drivers, Domino's discontinued its delivery guarantee.[2] Although its delivery service kept the company afloat, competitors were gradually closing the gap. Domino's hit a low point in 2008, marked by sluggish sales, plummeting share prices, and ongoing criticism of its pizza's flavor.

The advent of digital technology presented new opportunities for the Ann Arbor, Michigan–based company to enhance its strength in order fulfillment, and Domino's seized this opportunity. In 2008, Domino's unveiled an online ordering system, incorporating a pizza tracker that enabled customers to monitor the real-time status of their orders.[3] While many people considered the app to be a marketing gimmick, it had a profound impact on Domino's business.

From the consumer perspective, the app made the pizza preparation and delivery process more transparent. Essentially, Domino's gave customers insight into every detail of their orders. From the moment the order was placed, customers could track their pizza's progress, noting when the pizza was placed in the oven, when the food was packaged, when it left the store, and how close it was to arriving at the customer's home. With a promise of accuracy to within forty seconds, the app significantly reduced the anticipation and anxiety associated with waiting for their orders, thereby bolstering customer satisfaction and trust.[4] On the operational end, the tracker app necessitated a revamp of the company's systems to collect real-time data on pizzas and employee behavior, thereby allowing the company to measure employee productivity more accurately. Food delivery platforms could not replicate this level of transparency, as they lack visibility into the operations of the third-party restaurants they partner with.

Domino's also leveraged digital technologies to enhance the pizza ordering process and transform its omnichannel approach. Analysis indicates that Domino's digital ordering process enables around thirty-four million different ways to customize its pizza.[5] This flexibility helps satisfy various customers' palates, expanding the menu of classic flavors. Domino's stand-alone Zero Click app simplifies reordering by allowing users to save their favorite and customized pizzas for easy future ordering. After a ten-second countdown timer, their orders are automatically placed, requiring no clicks, swipes, or taps. Today, customers can place orders via more than fifteen channels, including chatbots, Facebook Messenger, tweeting a pizza emoji on X/Twitter, voice assistants like Amazon Alexa and Google Home, alongside the conventional phone, website, and mobile apps.

Dennis Maloney, the company's former chief digital officer, defined Domino's as a "delivery expert." "If everyone becomes an expert, we'd have to devise other ways to distinguish ourselves," he remarked in a 2017 interview.[6] The company made the foray into drone delivery and tested it out in the United Kingdom in 2013.[7] That year, then-CEO Patrick Doyle made IT Domino's largest department at headquarters.[8] The company's latest delivery innovation involved a partnership with robotics developer Nuro, facilitating autonomous pizza delivery via ground robots in Houston.[9]

To further strengthen its delivery capability, Domino's opened more stores closer to customers, thus reducing delivery time and ensuring fresher pizzas upon delivery. Between 2015 and 2020, nearly twelve hundred new restaurants were opened in the United States, with only eighty closures.[10] Food delivery platforms could not replicate this value proposition, as they do not own their partnered restaurants. In contrast, many delivery platforms took the opposite approach—expanding their delivery radius to accommodate more customers, making their platform more appealing to restaurants, but unavoidably lengthening delivery times.

Domino's, with its extensive delivery expertise, resisted partnering with food delivery platforms before and during the pandemic, unlike competitors including Pizza Hut and Papa Johns, who quickly formed such partnerships. Only recently has Domino's begun collaborating with Uber Eats. However, in contrast to many restaurants on the Uber platform, Domino's will continue to use its own drivers for pizza delivery instead of relying on Uber Eats drivers.[11] By concentrating on its strength in order fulfillment and leveraging digital technologies to amplify this strength, Domino's successfully bounced back from its low point in 2008.

Sephora: Building moats around personalization

Similar to Domino's, Sephora—the French multinational purveyor of beauty and personal care products—has also confronted tech giants and chose to double down its strength in personalization through digital innovation. Touted as a "retail apocalypse" survivor after battling with Amazon, the company, founded in 1970, has not only flourished in its brick-and-mortar retail presence but has also migrated its strength in personalization to the digital world.[12]

Personalization has always been important in the beauty industry as one customer's optimal makeup or skincare is rarely appropriate for another. Sephora pioneered the concept of try-before-you-buy for cosmetics, an innovation that has been widely replicated across the beauty retail sector. In its brick-and-mortar stores, Sephora staff members, known as beauty advisors, offer customers suggestions on products that may work best for them. It also offers in-store customized makeups and group classes.

However, with the growth of online beauty retail, driven by the advent of the internet and mobile e-commerce, the boom of social media marketing, and increasingly sophisticated consumers, digital-native companies pose a threat to Sephora. As early as 2023, Amazon entered the prestige beauty sector—a category between value and luxury that includes fragrances, skincare products, and cosmetics that are higher-priced than those sold at drugstores but less premium than high-end brands. It redesigned a new consumer interface that looks and feels like other prestige retailers. Customers can find unique product launches, highlights, editor's picks, sleek images, prominent brand logos, and category e-merchandising—similar to those found on Sephora.com. In March 2016, Amazon added a lifestyle show called "Style Code Live," which offered beauty tips, how-to's, and a live interactive chat feature.[13] The goal of Amazon Beauty was simple: offer consumers the same selection as other beauty retailers.[14]

In response, Sephora has innovated to enhance its customers' personalized shopping experiences by deploying digital technologies across web, mobile, and brick-and-mortar landscapes. One of the first digital tools Sephora introduced was the Pocket Contour app, in which users can upload photos of their faces to get a step-by-step primer on facial contouring using shading and highlighting. Users are then provided with product recommendations.

Another innovation Sephora instituted was its Virtual Artist app. Offering virtual makeup, the app delivers "an infinite library of eyeshadows, lip colors, and even false lashes to find your perfect shade and perfect your lip—all without stepping foot in a store."[15] The Virtual Artist, empowered by AI facial recognition technology, helps customers sample products digitally and try different shades of makeup after face-scanning via the app. If they like their simulated look, they can quickly and easily buy the products without leaving the app. These apps simplified the in-home cosmetics shopping experiences for customers who previously felt obliged to physically touch and sample products before purchase.

Sephora went a step further to amplify its strength in personalization from digital apps to online community building. In 2017, it launched a new online social platform, the Beauty Insider Community, which later became one of the world's largest online beauty forums. Shoppers engage with like-minded beauty experts, while enthusiasts and green hands virtually gather, ask questions, share makeup ideas, and connect through the platform. Unlike Amazon reviews, which offer limited and sometimes dubious feedback, the interactive community was designed for clients seeking a deeper level of beauty connection and inspiration from those whose preferences and insight they trust.

As more customers shopped across channels, Sephora focused on blending online and in-store experiences for a clearer view of customer behaviors. Like Domino's, the beauty retailer also adopted an omnichannel strategy by combining online and in-store teams to create an "omniretail" department. With this strategy, Sephora merged online and in-store customer profile data to help them make better purchasing decisions, thus driving sales.

Sephora has been quick to embrace tracking technologies, such as RFIDs (radio frequency identifications) and beacons that connect to in-store shoppers' mobile devices and guide them to their shopping choices on store shelves. Customers can scan any in-store shelf products with the Sephora app to consult online reviews and product ratings. The app provides access to a history of purchases, searches, and wish lists. It also retrieves scannable loyalty cards or saved gift cards, reducing checkout time.

While many retail brands struggled to build new customer points of interaction to bolster their brands during the Covid-19 pandemic, Sephora efficiently moved in-store personalized interactions into online ones that feel just as reciprocal. Despite store closures in many regions due to the pandemic, the company enjoyed a profitable 2020, with online sales breaking records globally. Sephora has topped the leading email marketing and marketing automation provider Sailthru's Annual Retail Personalization Index, which ranks 100 retail brands, every year since the survey first launched in 2017.[16] This recognition demonstrates the company's success in navigating both in-person and virtual formats, providing a shining example for other companies to follow as they establish a market position that tech giants find difficult to rival.

Avoid Becoming a Jack-of-All-Trades, Master of None

Conventional wisdom indicates the importance of agile adaptation in response to competition. Being agile, however, does not guarantee success. In fact, quickly following market trends or copying tech giants can sometimes be fatal.

Take the downfall of the Chinese retail conglomerate Suning Holdings Group as an example. Founded in 1990, the company started from a 2,153-square-foot store that sold air conditioners and grew to be China's once-dominant home appliance retailer with 22.6 percent of market share.[17]

With many opportunities created by digital technologies, Suning started an ambitious expansion. Founder Zhang Jindong set Suning's goal of becoming the "Walmart plus Amazon of China" in a speech at Stanford University in 2013. Suning entered almost every hotspot sector in China, where tech giants such as Alibaba also had a foothold. Over the years, Suning rose from an appliance retailer to a sprawling group encompassing sectors such as e-commerce, finance, sports, and entertainment.

The company's Suning.com site expanded from selling home appliances to various other goods. Suning established its finance unit in May 2015, a few months after Alibaba set up its fintech arm, Ant Financial (which later became Ant Group), in 2014.[18] While Alibaba's founder, Jack Ma, purchased half of China's then most successful soccer club Guangzhou Evergrande for 1.2 billion yuan ($192 million) in June 2014, Suning acquired another Chinese soccer club for 523 million yuan ($83.2 million) in 2015 and a controlling stake of Italian soccer club Inter Milan for 270 million euros ($322 million) in 2016.[19] These moves allowed Suning to match its competitors and tap into the health and lifestyle consumer market.

During the 2015–2019 period, the company's total investment reached 71.6 billion yuan ($10.7 billion), with high expectations of leveraging its new businesses to drive traffic for its core retail business.[20] However, Suning soon discovered that some of its new businesses were cash-burning, forcing it to subsidize them with cash flows from its retail business. For example, it made a substantial investment into broadcasting rights for European soccer league matches, which broadcast on its PPTV, a streaming platform acquired in 2013.[21] In the e-commerce space, Suning's market share was squeezed to just 1.7 percent in 2020, compared to the combined market share of 73 percent of the two market leaders in China, Alibaba and JD.com.[22]

Financial woes eventually hit Suning hard in 2021 after its offline retail business was severely dampened by Covid-19. The company's debt stood at more than $6.6 billion, with two-thirds of this amount being short-term obligations as of the third quarter of 2020. Digital titan Alibaba joined a local government-led consortium for a $1.36 billion bailout of Suning in July 2021.[23] The rescue plan also cost Chairman Zhang control and stewardship of the company he'd founded thirty-one years earlier, officially marking an end to the Suning era in China's retail industry.

Like Suning, numerous companies believe they need to diversify their businesses in the style of tech giants to thrive in the digital age. They've observed Amazon's journey: beginning with online book sales, expanding to a wider range of goods, and eventually developing various sectors such as on-demand video streaming, Amazon Web Services, hardware devices like Kindle and Amazon Echo devices, brick-and-mortar grocery stores, and more. Likewise, Google, initially a search engine provider, expanded both online and offline businesses, from software to hardware. The rapid, expansive growth of these digital superpowers might suggest that the key to success in the digital era lies in swift scope expansion.

However, these tech behemoths ventured into new territories only after constructing robust defenses around their core businesses—Amazon in e-commerce, Google with a search engine market share of over 70 percent in numerous countries. Traditional businesses must not underestimate the vital necessity of fortifying their strengths to ensure that their core businesses can withstand the onslaught of tech giants. Venturing into new territories typically demands substantial initial investments and often takes time to yield meaningful returns that can positively impact your bottom line. Without robust defenses to protect and develop your existing businesses, you might quickly lose ground to newcomers. Overshadowed by disruption in your core businesses and coupled with insufficient returns from new endeavors, your company could swiftly face financial distress.

Reflecting on the Sephora example, the beauty retailer could have sought to expand by branching out into sectors like clothing or jewelry after drawing a substantial user base fond of premium beauty products—a strategy many may recommend. Prioritizing diversification could be risky if these new ventures do not substantially enhance Sephora's strength in personalization. Instead, Sephora chose to concentrate on enhancing personalization within the beauty market.

How Do You Identify the Right Strength to Amplify?

Traditional businesses are built on a broad array of capabilities, which naturally raises the question: Which strength should you amplify? To address this question, it's necessary to scrutinize every key activity in your business and assess its relevance in the contemporary landscape.

One of us, before joining Harvard Business School, worked in the media industry, an example of how businesses have adapted within the rapidly changing digital environment. Print media, particularly newspapers, were often designated as a sunset industry due to the swift and vast progression of technology.

To illustrate, let's first analyze and unbundle the traditional business model of newspapers. In the old days, the value creation of such businesses centered on content creation, printing services, and a large network of delivery; the readership, in turn, allowed the newspapers to sell advertising space (figure 1-1). When technology eroded the previous business model of delivering ink-pressed newspapers to peoples' homes, content generation and advertisement attraction remained the primary value creation activities in the digital age.

However, the growth in targeted advertising, monetization of content, affiliate networks, and user data powered by tech titans like Google and Facebook have made advertising space in a prominent newspaper much less attractive.

Furthermore, sites like Craigslist, which offer free classified ad listing services, have significantly diminished the necessity of placing such ads in a newspaper. Ultimately, content creation emerged as the key value-creating activity for newspapers to fortify their competitive edge and distinguish themselves from digital players. High-quality content also helps attract and retain more advertisers.

FIGURE 1-1

Evolution of a newspaper's value creation activities

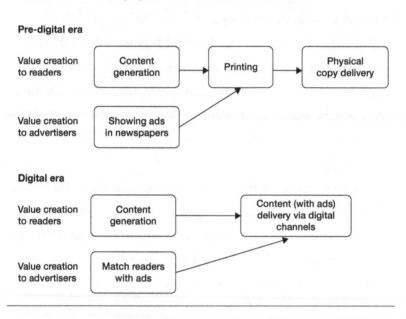

The New York Times may divide public views, but even people who disagree with its editorial stance must acknowledge that the nearly 180-year-old paper was able to strategically transition to the digital era. The *Times* is accelerating instead of slowing down, passing ten million paid subscriptions as of 2023.[24]

This growth has been driven partly by readers seeking out reliable news sources during the pandemic.[25] Born long before the digital age, the *Times* bet on readers' demand for high-quality, original, and independent journalism—a rarity amid the social media boom. As it became harder to capture value from advertising, the paper put up a digital paywall as early as 2011. With a newsroom staffed by seventeen hundred people and a strong presence in international reporting, it set a strategy of "digital-first, subscription-first" in 2015.[26]

It even disbanded its sports desk in 2023 to focus on news with "higher impact."[27] The virtuous circle goes: the more unique content the *Times* has, the more subscriptions it will acquire, and then the more investment it will be able to make in content.

The digital strategy paid off. Its digital-only subscription revenue surpassed print revenue for the first time in 2020 (figure 1-2). Subscription revenue of $1.4 billion accounted for two-thirds of the total revenue of $2 billion—making 2021 its first $2 billion year since 2012. In 2022, digital news accounted for close to 8.8 million subscribers, compared to about seven hundred thousand for the print newspaper.[28]

FIGURE 1-2

The New York Times **subscription revenue (in $ millions)**

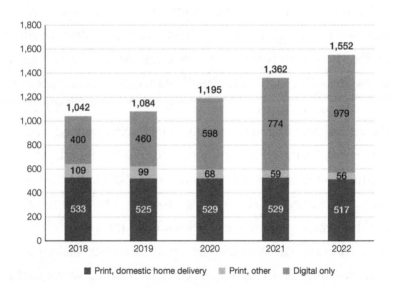

Source: Data from New York Times Company, 2022 Annual Report, https://nytco-assets.nytimes .com/2023/03/The-New-York-Times-Company-2022-Annual-Report.pdf.

Note: Print, domestic home delivery includes access to some digital products; print (other) includes single-copy, New York Times International and other subscription revenues.

A similar example is found in the world's largest furniture retailer IKEA, known for its brick-and-mortar stores since its founding in 1943.[29] Its physical stores served as a combination of showrooms, retail outlets, and immediate pickup points.

IKEA stores are also equipped with restaurants and child-care facilities designed to enhance the shopping experience and give adults some free time to explore. The Swedish furniture company has been known for providing free pencils and paper tapes for customers to measure furniture by themselves inside the stores. For many, a trip to IKEA is a fun day out for the whole family. However, during the pandemic, such value was significantly weakened. The rise of Amazon and other furniture e-commerce retailers, such as Wayfair, further reduced IKEA's value as a retailer in the United States.

In response, the company recognized that the value of showrooming remained relevant in the digital age. It doubled down on the showroom function by leveraging AI to offer customers an enhanced virtual home visualization experience. It acquired AI imaging startup Geomagical Labs in 2019. Functioning as the virtual pencils and tapes, the AI technology helped customers choose and perfect their home designs via mobile phones, scanning and transforming rooms into 3D models, allowing customers to try out IKEA home products. The tools mirrored the shopping experiences in brick-and-mortar stores.[30]

In addition, IKEA also empowered its instant pickup function by launching contactless Click & Collect services that allowed people to place orders via its app and collect products curbside. By emphasizing its strengths and creatively employing digital technologies to enhance them, IKEA endeavored to safeguard its market position against tech rivals.

Things to Look Out For

In this chapter we have emphasized the importance of reinforcing and amplifying your strength using digital technologies to compete effectively against tech behemoths.

It's essential to remember that strengthening your competitive edge is not solely about improving the execution of your existing strategies. Rather, it involves leveraging technology, business model innovation, and other emerging opportunities to redefine your business. For instance, luxury car manufacturers like Ferrari should contemplate how technology can be used not only for more efficient production but also to heighten the luxury experience for car buyers in a manner that tech behemoths can't replicate.

Amplifying your strength also does not mean that you should avoid diversification. For instance, Domino's autonomous delivery experiments, if successful, hold the potential to either scale or evolve into a separate delivery enterprise for other restaurants.

Disney, another example, was established well before the digital era but has employed technology to captivate and engage customers beyond movie production, extending to theme parks and physical stores. Since its inception in 1923, Disney has expanded into businesses closely aligned with the company's core strength of unmatched storytelling capabilities, both onscreen and in-person, and the intellectual property rights of thousands of characters and movies. To compete with tech giants like Amazon and Netflix, Disney abandoned its licensing strategy, regained control of its own content, and launched the video streaming platform Disney+ in 2019.[31]

Disney+ offers movies and videos from Pixar, Marvel, and 21st Century Fox, thanks to a series of acquisitions that may have seemed expensive at the time. Although the platform's content library is considerably smaller than that of rivals such as Netflix or Amazon, its strengths lie in quality over quantity and an array of evergreen content. The direct-to-consumer service allows Disney to gain better control over its customers' experience and reinforce its reputation for producing high-quality, family-friendly entertainment. By the fiscal quarter ending September 30, 2023, Disney+ had topped 150 million subscribers.[32]

As companies actively strive to amplify their strength in the digital era, numerous appealing diversification opportunities might naturally come to light. In subsequent chapters, we will illustrate these opportunities with additional examples.

CHAPTER 2

Drive Customer Centricity

We have always wanted to be Earth's most customer-centric company. We won't change that.

—Jeff Bezos

Brick-and-mortar stores are often viewed as a relic of retail's past rather than its future. However, one Chinese shoemaker begs to differ.

This retailer, which collected data on shoe try-on rates within its stores, inserted smart chips into its footwear. Surprisingly, the data revealed that the most frequently tried-on shoe type didn't top the sales charts. Upon discovering and redesigning excessively long shoelaces, the conversion rate from try-ons to purchases for this specific shoe type soared from 3 percent to 20 percent. In addition, the retailer installed 3D foot scanners in over one thousand of its retail locations across China, providing complimentary foot measurements for customers. Within thirty seconds after a customer steps on the scanner, a measurement and a personalized shoe recommendation are completed.[1]

Enter Belle, the leading retailer of women's footwear in China by revenue. Like many traditional businesses, Belle, established in 1992, grappled with a scarcity of customer data to guide product improvements and elevate customer satisfaction.[2] As of 2021, Belle had amassed foot-shape data from roughly four million individuals, the most extensive database of its kind in China.

Unlike tech giants, such as Alibaba, which make product recommendations based on the data of customer feedback and purchasing history, Belle's data collection approach takes its recommendation system beyond style and price preference—fit and comfort are two important elements often hard to capture online. It also helped the company shorten its research and development time, and offer customers better choices, especially those with less common foot sizes.[3] Belle's experience shows that traditional businesses can come up with their own approaches to collect data that tech giants cannot easily imitate and drive their customer centricity in the digital age.

Tech Giants' Approach to Customer Centricity

The notion of customer centricity was popularized long before the advent of tech giants. As early as 1954, Peter Drucker, often considered the founder of modern management, stated in his book *The Practice of Management* that "it is the customer who determines what a business is, what it produces, and whether it will prosper."[4] In the service industry, the mantra "the customer is always right" has been widely adopted since the start of the twentieth century.[5] Variations of this motto in German and Japanese analogized the customer to either a king or a god.

Today, with the onset of digitization, data has emerged as a pivotal asset in driving customer centricity.

Tech giants are often acclaimed as playing the customer-centricity strategy to the fullest. Consider Google's search engine: every user search and subsequent click on a particular link among displayed results helps refine Google's search ranking algorithm, attracting more users and customers. This self-reinforcing cycle, known as data network effects or the data flywheel (figure 2-1), when strong enough, can make their business highly attractive, leading to a winner-takes-most or even a winner-takes-all dynamic. Amazon's and Netflix's recommendation systems, along with many large language models today, also take advantage of such data network effects. The more customers use the service, the better the recommendations or results offered by service providers, and vice versa. Netflix even employs data insights to produce films, directly challenging

FIGURE 2-1

Data flywheel

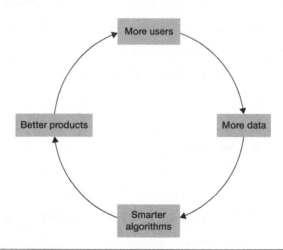

Hollywood. Many viewers find Netflix originals more enticing than traditional Hollywood productions.

Without the same scale of data collection and algorithm processing capabilities, most traditional companies are worried about losing customers to their tech rivals. Despite substantial investments in digital channels and attempts to collect and analyze data like tech giants, they often find themselves falling further behind due to the exponential growth of these technology behemoths. Indeed, an increasing number of brick-and-mortar stores have either shuttered or are on the verge of closure. At the same time, just like Belle, instead of merely imitating tech giants, many traditional companies have also found their secret recipes, approaching data collection in their own ways to drive customer centricity.

The Four Pillars of Customer Centricity

Customer centricity contains four pillars: product improvement, simplifying customer journey, personalization, and satisfying customers' real needs (figure 2-2). In this chapter we discuss the strategies traditional businesses can employ to overcome data paucity in each pillar.

Product improvement

To be customer centric, companies need to drive product design and improvement based on consumer insights. Tech giants enhance their offerings by tracking real-time customer data flows.

FIGURE 2-2

33 Drive Customer Centricity

Four pillars of customer centricity

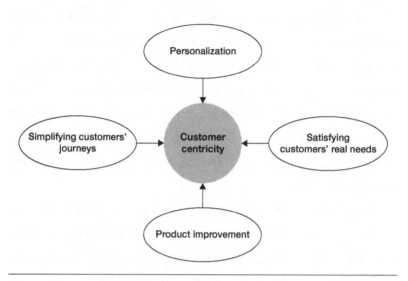

Traditional firms may not have the same level of data access in their digital channels, but they often can, similar to Belle's strategy, utilize their existing offline channels to complement their digital channels to glean useful insights. This dual approach can provide critical insights for improving their products or services.

Coca-Cola: Harnessing offline data for product development. Coca-Cola, the global beverage producer with a history of over 130 years, was the world's strongest brand among all the nontech brands in 2023, trailing five tech giants, Apple, Google, Microsoft, Amazon, and Facebook.[6] To drive its brand recognition, in addition to leveraging online channels, Coca-Cola enhanced its product offering through customer data collected from offline interactions.

In its offline operations, Coca-Cola typically interacts directly with distributors and retailers. However, the company has identified a direct consumer connection through its self-service Freestyle soda fountains. In 2017, after analyzing real-time data garnered from those soda fountains, Coca-Cola introduced Cherry Sprite as a new permanent flavor. These touch screen soda fountains can dispense over 100 different flavors, enabling customers to blend flavors on their app and pour from the machine.[7] Studying how customers interact with these fountains provides Coca-Cola with valuable insights on customer preferences and trends, making its new offerings more successful.

To facilitate more interactions like this, in 2019 Coca-Cola initiated its first "Make Your Mix" contest, encouraging users to personalize their drinks. The general public then voted on the most popular recipes that contestants posted on social media.[8]

Such data gathering efforts represent a win-win for Coca-Cola and its customers. Often, customers find it personally rewarding to contribute to the development of products they value, deriving satisfaction from their role in improving or popularizing these products. Many are willing to offer their insights for free. With digitalization, more products are being enhanced by internet of things (IoT) technology. This development enables firms to connect with their consumers through offline channels in real time. Therefore, traditional businesses should proactively utilize their offline channels to involve customers in product development. Emulating Coca-Cola's approach, they should facilitate customer experimentation with their products and derive insights from these engagements.

Patagonia: Leveraging data for sustainability. Patagonia, a global leader in environmentally responsible business practices and an outdoor apparel retailer with annual sales exceeding $1 billion, leverages insights from customer product return data to drive product improvement. Patagonia's diverse portfolio spans approximately fourteen hundred products, including fleece jackets, casual T-shirts, high-performance athletic and outdoor wear, camping gear, backpacks, and wetsuits.[9]

In 2017, as a sustainability pioneer within the fashion industry, Patagonia launched the Worn Wear program online.[10] This program was created to encourage customers to repair, recycle, and reuse their items, instead of disposing of used gear and buying new products. Items in good condition could be returned for credits, and these used products would then be repaired and resold on Patagonia's Worn Wear website. Supported by over seventy repair centers worldwide, the company's Reno, Nevada, facility repairs more than one hundred thousand items each year.[11]

Data from this program has been instrumental in enabling Patagonia to improve its product quality and meet its sustainability goals. This data helps the company monitor reselling activities and the lifespan of resold items. For instance, if a product is returned significantly earlier than the average five-to seven-year lifespan, Patagonia might consider modifying the product's design or discontinuing its production altogether.[12] If an item is resold multiple times, it prompts the design team to explore and address any potential design issues. Using this data, the company fine-tunes its manufacturing and sales strategies for upcoming products. Likewise, when items are frequently returned for repairs, the company considers design improvements to rectify recurring issues.

The company's efforts began to bear fruit as it sold over 120,000 repurposed items in just over two years since the program's inception.[13] For consumers, not only could they feel good about their purchase of a more sustainable product, but the knowledge that they could return the product for credits later also incentivized initial purchases.

As the stories of Belle, Coca-Cola, and Patagonia show, customer data regarding what they viewed, purchased, or returned is valuable to both digital natives and traditional companies for product improvement. Traditional businesses need to think outside the box to identify novel data collection opportunities that tech giants find difficult to grasp. Their existing channels could serve as sweet spots. In the United States, e-commerce accounted for about 15 percent of the total retail industry in 2023, indicating that most transactions still took place in brick-and-mortar stores, the stronghold of traditional sellers.[14]

Simplifying customer journeys

Leveraging AI and data, tech giants have substantially simplified customer journeys, providing seamless experiences. For instance, MYbank, a subsidiary of Chinese fintech titan Ant Group, is among the country's first digital-only banks. It's best known for its "310" lending operation model: it provides round-the-clock small loan approvals, promises user registration in three minutes, loan approval in one second, and zero human intervention.[15]

MYbank primarily serves small and medium enterprise (SME) owners and farmers, who can conveniently apply for microloans via its mobile app.[16] These borrowers are not attractive to traditional banks because the majority of them don't have credit history. MYbank minimizes the risk by leveraging data from its parent company Ant Group's e-wallet payment platform Alipay and Alibaba Group's e-commerce sites, which have over 900 million users and one billion annual active users in China respectively.[17]

Taking a data-driven approach, the bank leverages AI to assess a borrower's payment ability.[18] Algorithms quickly evaluate the SME's repayment patterns, and the owner's online shopping history, which then automatically sets credit limits and interest rates.[19] These risk management technologies helped hold the bank's nonperforming loan ratio at 1.52 percent, significantly lower than the national average of 2.99 percent for SME loans as of June 2020.[20]

Posing a threat to the traditional banking model, MYbank simplifies the customer journey considerably thanks to its access to massive data about each customer that allows it to assess a customer's risk efficiently and accurately (figure 2-3).

FIGURE 2-3

The customer journey of loan applications

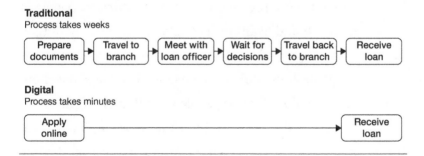

Repeatedly, we observe that tech giants' strategy to enhance customer satisfaction isn't merely about digitizing the traditional customer journey, but rather about utilizing available data and AI to significantly simplify it. MYbank is just one example. Numerous e-commerce titans are employing innovative methods to simplify the customer journey via data and AI. For instance, in the future retail landscape envisaged by China's leading retailer JD.com, if a consumer sees a nice bag on the street, she could scan it with her phone's camera (with the owner's permission), and an augmented reality app would take her directly to an online store to make a purchase.

Traditional loan and insurance companies have long understood that offering instant quotes to their customers significantly aids in both attracting and retaining them. The risk of potential customers being poached by rival companies escalates as the time taken to follow up lengthens. However, the question remains: How can these companies provide real-time, reasonable quotes akin to those offered by MYbank, especially in the absence of extensive and precise customer data?

Ping An: AI and behavioral insights. In China, Ping An Group, a veteran insurance company, has employed behavioral economics to simplify the loan application process for its customers.[21] Ping An grew from a regional insurer with thirteen people in 1988 to a fintech and healthtech conglomerate with more than three hundred thousand employees and nearly 1.4 million insurance sales agents at its peak, which means that one in one thousand of the 1.4 billion Chinese people were Ping An agents.

While Ping An did not have the same types of data as Ant Group and JD.com, it developed the Micro-Expression Remote

Interview System and AI-powered Sentiment System to help its own subsidiaries and other financial institutions identify early signs of fraudulent behaviors when they interview customers for loan applications.[22] The real-time analysis of thirty-nine facial actions, including rapid blinking and eyeball movements, recognizes attempts to suppress emotions. The company found that nearly 90 percent of loan approval decisions made based on this system were consistent with those made by humans, cutting labor costs by about 40 percent and avoiding human-made pitfalls.[23]

Numerous traditional businesses beyond the financial sector have also devised innovative strategies to simplify the customer journey. In the hospitality industry, for example, both Marriott and Hilton Hotels have introduced a mobile key feature in their apps. Guests can check in, unlock their rooms, and check out using just their smartphones, bypassing the traditional front desk visit and saving significant time. Such innovations encourage travelers to use the hotel apps, fostering loyalty and enhancing the hotels' competitiveness against digital-native platforms like Airbnb and online travel agencies like Booking.com.

Retail giant Walmart, in its bid to compete with Amazon's seamless online shopping experience, has also integrated technology into various aspects of its in-store shopping journey. The Walmart mobile app, for instance, features a map customized to each of its store locations, guiding customers to their desired items. Moreover, its Scan & Go technology enables customers to scan items while shopping and pay via the app, thus eliminating the need to wait in line at the checkout.

Returning to the example of Domino's from chapter 1, the pizza company's stand-alone Zero Click app makes reordering incredibly straightforward, with users able to save their favorite and customized pizzas for future instantaneous orders. The key piece of data Domino's requests from its customers is simply their favorite type of pizza. As food delivery platforms offer a vast variety of food choices, they may find it challenging to replicate Domino's customer journey simplification.

As these examples demonstrate, simplifying the customer journey goes beyond reducing the number of steps in the journey; it's about creating a more enjoyable, accessible, and engaging experience that puts the customers' needs and preferences at the forefront. These examples also show that simplifying the customer journey doesn't always require massive amounts of data as one might expect. Even in this data-driven era, traditional businesses can employ creative strategies to revolutionize their customer journeys.

Personalization

In the digital age, companies have more opportunities to know their customers better. Extreme personalization, although very difficult to achieve, aims at leveraging AI-related technology to offer the right product to the right customer segment at the right time.

Today, the world's largest digital advertisers are Google, Facebook, and Amazon, accounting for more than 63 percent of all US digital ad spending since 2019.[24] They can target customers efficiently based on a large mass of data. They know where the customers live, what they buy, watch, read, and more. Companies must pay the toll to use their channels to target the right customers.[25] For example, some third-party sellers are paying more than $100,000 a year to Amazon for its targeted advertising service.[26]

Traditional businesses can adopt a distinct approach to personalization. First, they can utilize their own channels for data collection and analysis and to deliver personalized offerings to their customers. Second, their deep understanding of their products or services allows them to tailor their personalization strategies specifically to each offering.

Coca-Cola: Make the best of your own channels. Coca-Cola was among the pioneers in leveraging its own channels, combined with an in-depth understanding of its products, to gain insights into customer preferences.

The Atlanta-based company, with over 1.1 million followers on the platform X/Twitter, 2.9 million followers on Instagram, and more than 109 million likes of the Coke Facebook page, closely tracks its products across social media platforms. With data-mining tools, the company not only monitors the effectiveness of its own posts across the internet, but also analyzes company mentions in the posts by its consumers. The company uses AI to understand how its customers or potential customers are discussing and engaging with the portfolio of Coca-Cola brands. By analyzing such extensive data, the company has gained insight into its customers' identity, location, and preferences.[27] Instead of randomly placing a commercial, Coca-Cola targets its ads to those customers with the highest likelihood of engagement.

The company also uses AI image-recognition technology to identify and analyze the photos of its own or its rivals' products by social media users. For example, it can spot the images that feature glasses of tea, or bottles of Snapple, Honest Tea, or Lipton with users of happy emotions on Instagram or X/Twitter. Based on these socially shared photos, Coca-Cola gains insight into these tea lovers: where they are from, and how and why the brand is mentioned; and, more importantly, it helps the company identify the right group of potential customers who might want to try new tea products.[28]

The company pushes targeted ads for its own Gold Peak Tea to them, across forty other mobile sites and apps, broadening the scope for where customers might come across them—for example, when reading an article on Business Insider or checking the AccuWeather app.[29] The click-through rate is four times higher than any other targeted advertising methods, according to Coca-Cola.[30]

Many companies are sitting on a gold mine of massive available public data, but they do not take advantage of them. Some companies attempt to use these resources but limit themselves to basic analysis, such as simply tracking and comparing the number of mentions of their and competitors' brands in text. However, advanced machine learning techniques now enable firms to analyze the photos and videos increasingly shared by users online, identifying patterns and gleaning deeper insights.

Kroger: Personalizing shopping carts. Unlike the asset-light business model employed by many tech giants, traditional businesses often maintain significant physical asset ownership. These assets are typically viewed as burdensome and a primary reason for the inability of traditional businesses to compete effectively with tech giants. However, models involving heavy asset ownership often lead to richer customer interactions. When leveraged effectively, these assets can enable the collection of more granular data, which tech giants may lack access to, thereby driving enhanced personalization efforts.

The US's largest supermarket chain, 140-year-old Kroger, developed its personalization strategy inside its twenty-eight hundred physical nationwide stores.[31] In the pre-internet days, supermarket promotions were routinely advertised in paper-based catalogs or through broadcast media. Today, though, Kroger has adopted infrared and IoT technologies with video analytics to track customers' shopping activities in its stores. It established an internal data analytic group in 2015, where it monitored the data collected from the store aisles and ran machine learning algorithms for sales forecasting and customer behavior assessment.[32]

Kroger's strategy didn't stay offline-only; it integrated the stores with its mobile customer interface. Like the Sephora example in chapter 1, Kroger's mobile app, when activated by shoppers and detected by in-store sensors, can send shoppers real-time personalized product suggestions, personal pricing, and digital coupons, based on their shopping and viewing history.

"Many retailers have transactional data, but no one has the customer data and the insights that Kroger has," said CEO Rodney McMullen on the company's 2020 third-quarter earnings call, during which he mentioned the words "personalization" seven times and "personalized" twice.[33]

Effective personalization in the digital age entails a sophisticated combination of data collection, analysis, and the application of insights to customize experiences, products, or services according to individual customer preferences. The experiences of Coca-Cola and Kroger illustrate that traditional businesses can adapt their personalization strategies to fit the distinct nature of their products and services, as well as their unique operational models. In contrast, tech giants typically employ the same data-driven strategies across a diverse range of products. While they may not need to develop digital infrastructures as comprehensive as those of tech giants, traditional businesses should invest in the necessary digitized operational backbones. This investment enables them to effortlessly, naturally, and instantly collect data and generate insights, which are vital for achieving effective personalization.

Satisfying customers' real needs

Data analytics today helps predict consumer behavior, but it has limits in understanding and revealing consumer motivation. While data gathering via human communications and interactions, slow and costly, may sound like dinosaur-age practice, it still shines like a diamond. Personal interactions help in building trust and rapport. When customers feel understood and valued, they are more likely to open up about their true needs, preferences, and pain points. Such interactions are thus highly valuable when it comes to understanding customers' real needs.

As many scholars have argued, understanding what customers really hope to accomplish is key to successful innovation and creating offerings that people truly want to buy.[34]

DBS: Human resources might be more resourceful than you think. Founded in 1968, DBS Group, Southeast Asia's largest bank and one of the country's oldest lenders, makes the case that traditional banks can delight their customers in their own ways. It was recognized as "World's Best Bank" and "World's Best Digital Bank" for 2021 by industry publication *Euromoney*.[35] The bank coined the acronym GANDALF— also a wizard in the *Lord of the Rings*—sandwiching itself in the center of the global tech giants **Google**, **Amazon**, **Netflix**, **Apple**, **LinkedIn**, and **Facebook**.[36] Yet, instead of imitating the big techs, DBS fully utilizes its own expertise and found its unique way to drive customer centricity.

While many tech giants, such as Ant Group, have a competitive advantage over traditional banks on technology deployment, they often get to know their customers only through cold big data and analytics. DBS has a critical asset that those tech giants don't have—frontline banking employees, who are also its point of contact to know customers better with a human touch. In addition to empowering them with technology, the bank trains them in active listening and empathy to ensure they can understand the underlying needs and concerns of clients.

Where most commercial banks offer mortgage services, DBS went beyond mortgage sales. Through their employees' interactions with their customers, they realized that the real need of many of their mortgage borrowers was to become first-time homeowners; therefore, the real goal for the bank should be to help them attain such dreams. Leveraging the trust these employees built with the customers, the bank thus set a goal to be part of their customer's complete home-buying experience and assisted them in home searching, mortgage shopping, and assessment—a process that would start six months earlier than a normal mortgage service.[37]

DBS took its advantage a step further, launching Southeast Asia's largest bank-led property marketplace in 2018—the first bank in the region to set up such a platform.[38] This one-stop shop allowed home buyers or renters to search around one hundred thousand listings, complete transactions, and apply for property, all on the same platform.[39]

The role of the bank would seem to be invisible to customers during their purchase journey, as they would feel like they were buying the home. The bank creates value by offering a home financial planner to smooth out the journey of first-time home buyers' affordability assessment. In fact, prior to entering the home market, DBS launched Singapore's largest direct seller-to-buyer online car marketplace and an electricity platform in 2017 with the same mindset.[40] In this example, even though customers plan to buy services from banks, their real need is outside the financial sector.

In 2018, DBS rolled out a "Live more, Bank less" rebranded tagline, as its research found that their customers' real needs were livelihood, while banking was just a means.[41] "Bank less" also meant that DBS embedded banking in the customer journey to help them accomplish their real needs, be it online or offline. The bank reported a record net profit of S$8.2 billion ($6.2 billion) in 2022, a 20 percent jump from a year earlier.[42] Share prices rose 4 percent in 2022, compared to other leading banks, which lost about 20 percent to 30 percent, including JPMorgan and Bank of America, in the same period.

Digital technology, with its myriad of benefits in efficiency and cost reduction, nonetheless encounters significant limitations in fully comprehending the complex landscape of consumer needs. The human touch offers empathy, understanding, and the ability to connect on a more personal level, which are critical in discerning and addressing the more subtle and real needs of consumers. In an era where digital technology is ubiquitous, genuine human interactions have become a rarity, thereby increasing their value in the business world.

Many traditional businesses today are inclined to prioritize technological solutions over human resources, aiming to enhance efficiency and reduce operational costs. However, this approach can inadvertently neglect the unique advantages that human interactions bring. When technology is used to complement and augment human interactions rather than replace them, a synergy is created where technology enhances human capabilities, leading to more creative problem-solving and a true understanding of customers' needs, and driving new value creation opportunities.

Nike: From shoemaker to lifestyle builder. Like DBS, Nike's focus on addressing its customers' real needs has unveiled many new growth opportunities. With footwear as its most valuable segment, Nike's value creation went far beyond manufacturing.[43] As the world's largest sneaker maker, Nike discovered that consumers seek more than just functional attire; they desire healthier lifestyles.

For example, in California, Nike introduced boutique training studios in 2023 that offer live and social fitness classes led by Nike trainers.[44] The studios, transcending Nike's iconic slogan from "Just do it" to "Let's do this," foster a fitness community where customers train alongside both old and new friends.[45]

Digitally, Nike has developed a variety of Nike+ mobile applications, forming a digital health ecosystem that spans training classes, nutrition guidance, and motivational support. These in-house-developed apps, including Nike Run Club and Nike Training Club, enable Nike to collect data from every customer interaction, informing the design of new products and services that promote a healthier, more active lifestyle. This approach resonates particularly well with the digital-native generation.

Moreover, Nike doesn't just respond to existing customer needs but also aims to shape and elevate the aspirations of potential customers. In January 2023, Nike launched thirty hours of fitness content on Netflix, available in ten languages.[46] This initiative targeted a segment of Netflix users who had not yet prioritized fitness, intending to encourage these individuals to become more active. Nike's emphasis on promoting fitness has a lasting impact on customer acquisition and retention because fitness becomes a routine and a regimen.

While tech companies often expand into multiple markets using data and large user bases, the stories of DBS and Nike demonstrate that traditional businesses can also expand by discovering and understanding customers' real needs. When customers feel understood and their needs are met, they are more likely to develop long-term brand loyalty. By truly understanding and meeting consumer needs with a diverse product portfolio, businesses can set themselves apart in a competitive marketplace. This differentiation is not just about offering products or services that are distinct from those of tech giants, but also about creating experiences that deeply resonate with consumers on a personal level.

Things to Look Out For

In this chapter, we've explored a range of strategies to tackle data scarcity and enhance customer centricity. The examples above underscore the necessity for traditional businesses to leverage their unique assets and capabilities to bolster customer centricity. Thanks to the rich, frequently human-touch interactions inherent in their operating models, and their comprehensive knowledge of their industries, traditional businesses are uniquely positioned to foster customer centricity in ways that tech giants may find challenging to replicate.

It's also crucial to understand the limit to big data. Let's look beyond the spinning data flywheel and examine its limitations to understand why or when data might not offer sustainable competitive advantages.

First, conventional wisdom might suggest that having more data is preferable, but in the majority of cases, the most significant gains from data can be achieved with a relatively small data set. The value increases noticeably with volume only up to a certain point: the first extra $10x$ of information is highly valuable, but the progression from $100x$ to $110x$ barely adds significant value (figure 2-4). This observation has been demonstrated in Netflix's and Amazon's recommendation algorithms, where improvements in users' likelihood to click on recommended items rapidly decline "beyond a modest threshold."[47] This is also evident in news personalization, where performance plateaus quickly with the addition of extra data.[48]

FIGURE 2-4

Volume of data and value creation

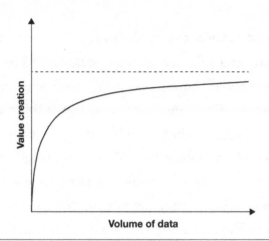

Second, the significance of edge cases—those that are atypical or uncommon—varies across different applications. Edge cases are crucial for systems designed to predict rare events, like diseases, or for search engines. For instance, in the case of Google's search engine, keywords queried less than ten times per month accounted for nearly 80 percent of its searches.[49] You could thus imagine that Microsoft's Bing or Yahoo's search engine would observe these keywords with much lower frequency, making it hard for them to match Google's service quality.

However, in many scenarios, the impact of edge cases is limited and manageable. Take the insurance policy market as an example again. Most applicants are typical users, allowing firms with small datasets to provide a service quality comparable to tech giants. Firms can address edge cases through human intervention, or in certain scenarios, choose not to serve these customers. This differs from the search engine market, where edge cases constitute the majority of searches and users expect immediate results.

Third, the value of data can be influenced by the presence of alternative data capable of achieving similar or even superior results—all roads lead to Rome. In many cases, a variety of data can predict the same outcome. For example, Ping An, in addition to the use of its data, takes advantage of behavioral economics as a complementary approach: users' behavior during their application processes can serve as a strong indicator for their risk profile. As for Coca-Cola, its own approach yields better performance than common methods such as purchasing keywords on Google or targeted advertising on Facebook.

For traditional businesses, understanding the limitations of big data is crucial because you may not possess the resources to collect and process data like your tech competitors. In the meantime, you should innovate to achieve customer centricity in your own way by leveraging your unique channels and resources. Stay focused and press on!

CHAPTER 3

Find a Platform Opportunity

E stablished in 2016, Zé Delivery (Zé) has evolved into one of Brazil's most valuable digital start-ups, specializing in last-mile delivery of beer and other beverages, with its core value proposition being "fast, cold, and affordable."

Boasting an average delivery time of twenty-two minutes, Zé ensures customers enjoy cold beer at fair prices. Zé was developed within Ambev, a subsidiary of Anheuser-Busch InBev (ABI), and the largest brewing company in Brazil and Latin America, which commands over 60 percent of Brazil's beer market. However, Ambev's growth trajectory began to decelerate in the mid-2010s due to competition with rivals such as Heineken. Its board was resolute to not only counter the current wave of competition, but to lead and take charge in the next one (e-commerce).

Zé became integral to its parent company's strategy to connect directly with beer consumers in Brazil. Zé facilitated the sale and delivery of not only Ambev's products but also those of third-party brands, including competitors such as Heineken. By 2022, the startup boasted a valuation exceeding $1 billion.

The Zé mobile app uses an algorithm to assign customer orders to beverage sellers—including bars, restaurants, small stores, liquor shops, and convenience stores—based on factors such as demand, inventory, location, and courier availability.

These sellers undertake the responsibilities of product stocking, delivery, and invoicing. To mitigate costs, Zé offers operational incentives and delivery subsidies, assuring sellers of a reasonable markup. Zé collects online payments and transfers them to sellers. Additionally, it operates its own warehouses and dark stores, enabling the platform to fulfill orders independently. Zé monetized its business by taking a commission on the price charged to consumers. The direct-to-consumer model offers Ambev the opportunity to gain rich consumer insights, which subsequently enable it to conduct effective target marketing.

Zé experienced a dramatic increase in order volume during the Covid-19 pandemic, and the company's momentum remained unabated even as lockdown restrictions were lifted. In 2022, Zé operated across all 27 Brazilian states, delivering 62 million orders to over eight million users through its app. With Zé, ABI—a traditional beer manufacturer—ventured into a platform business. Zé's journey demonstrates how traditional businesses can discover platform opportunities within their own operations and scale such businesses based on their unique resources and capabilities.

Paths to Platforms

Platforms, which have been prevalent in various forms throughout history—from matchmaking services to medieval Italian fairs for agricultural exchanges—are intermediaries that link multiple user groups and enable direct interactions. Modern tech giants like Apple, Microsoft, Alphabet (Google), and Amazon generate and capture considerable value through their platform business models. Food or beverage delivery platforms typically operate on a three-sided model (connecting three user groups: merchants/restaurants, consumers, and couriers).

Zé's unique model, with most of its sales originating from its parent company's products, evolved into a four-sided business model that links Ambev, consumers, sellers, and couriers (figure 3-1). Zé determines prices, take rates, and subsidies, ensuring a reasonable margin for sellers and affordable beer prices for consumers. With a beer gross margin of up to 60 percent, Ambev was capable of investing in Zé to support this pricing strategy. This effectively addressed the challenge of attracting both merchants and consumers, fueling the platform's growth.

Many Zé customers discovered that beer ordered for home delivery was not only cheaper but also more likely to stay cold

FIGURE 3-1

Zé Delivery's business model

Source: Adapted from Ambev Investor Relations, "Presentations, Investor Day: 9. Zé Delivery," April 12, 2022, https://api.mziq.com/mzfilemanager/v2/d/c8182463-4b7e-408c-9d0f-42797662435e/59e92e52 -0ff0-4967-5bd9-dca04a4a14f3?origin=1.

upon arrival than if they had made a trip to the store. Zé also expanded its offerings to other fast-moving consumer goods, such as snacks and barbecue products, to become a platform that enabled social occasions. The platform created value for Ambev, providing consumer data and insights, establishing a new channel for the low-cost development of beverage brands, and introducing new monetization opportunities such as advertising. Most significantly, it drove the sales of core beer products.

Companies typically become platforms via one of two approaches (figure 3-2). Some businesses start off by

FIGURE 3-2

Two approaches to becoming platforms

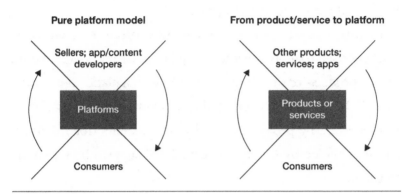

launching an intermediary that allows various groups to connect, interact, and transact; eBay, which enables sellers and buyers to transact directly, serves as an illustration of this model. With this approach, platforms often leverage cross-side network effects, where the two sides attract each other, to grow.

On the other hand, many companies, including a majority of tech giants, undergo a product-to-platform transformation similar to Zé. For instance, Google started as a search engine in the mid-1990s and later expanded into search advertising to link advertisers with users. Apple created the iPod in 2001 and subsequently developed the iTunes Store and the App Store. Amazon, having initially achieved success in retail, launched its marketplace in 2000 to work with third-party sellers.

In a similar vein, OpenAI expanded its ChatGPT service to incorporate third-party plug-ins after amassing a user base of over 100 million. These plug-ins boost the capabilities of

ChatGPT, empowering it to perform a broad array of actions, including fetching real-time information like sports scores, booking flights, and ordering food.

In this second approach, businesses often leverage their existing user base as a launching pad for their transformation from a product- or service-oriented model to a platform-oriented model. In addition to cross-side network effects, platforms attract users by providing stand-alone benefits (for instance, search service by Google or ChatGPT service by OpenAI) to them.

Keys to Success

What does it take for a traditional business to find a platform opportunity in its own business and nurture its growth like Ambev? After over a decade of researching the transformative journeys of various traditional businesses, we've identified several critical factors for success.

Developing an open mindset

It takes courage and time to shift mindsets from product to platform for any product-centric firms. Embracing third parties is often easier said than done—even for Apple and Steve Jobs. Apple celebrated the sixteenth anniversary of its App Store in July 2024.[1] We are old enough to recall that the first iPhone just looked similar to an advanced personal digital assistant (PDA) at the time. It was a closed-system product without an app store. The Jailbreak Community was then born and blossomed as hackers began to develop and install their own apps on the iPhone. Then-CEO Steve Jobs's knee-jerk reaction was to go on the defensive. He initially wanted to make the operating system more secure and threatened to punish the jailbreakers, including by voiding the warranty for anyone who hacked the device.

However, the third-party apps showcased the value creation opportunities that an iPhone could have. Jobs and Apple came to see the opportunity of creating a more open platform and officially opened the App Store in 2008, a year after launching the iPhone.

The simple digital storefront reached one billion downloads within nine months.[2] Apple also captured value from the App Store by taking a cut between 15 percent and 30 percent of app sales and in-app purchases.[3]

Similarly, LEGO initially resisted incorporating outside designs, preferring to keep product development in-house. However, it eventually adopted a more open mindset by embracing fan involvement. A notable example of this shift is the LEGO Ideas Platform, which enables fans to submit designs for new LEGO sets. When a design gains sufficient community support, it undergoes a review process by LEGO and it may become a commercial product. This approach expanded LEGO's product range and deepened its fan community ties, resulting in popular sets like the NASA Apollo Saturn V.[4]

Valve and Steam: Working with hackers. Another example of a product company adding a platform is US game developer Valve Corporation, known for working with hackers. By trial and error over the years, it learned an open platform would be an optimal solution to deal with hackers as well as create value for users.

Founded in 1996, the company developed its first video game, *Half-Life*, which was so popular that hackers almost immediately started cracking and modifying the game. Unauthorized modifications often made the game unstable and prevented users from playing together on a stable network. Instead of fighting off the onslaught of hacking, Valve chose a different route by working with the very hackers who had developed the most popular mod. The unmet demand for alternatives also offered new value creation potential for the company. Valve asked the hackers to develop a new game, *Counter-Strike*, which also proved wildly popular. However, hackers continued to crack the games that created significant problems for gamers. In 2003, Valve released an online software channel called Steam where it distributed patches to gamers. In 2018, the company paid $20,000 to a hacker who discovered a critical security bug on Steam.[5]

As the number of users on Steam grew, the company transformed it into an online platform to solve its distribution challenge: By letting users download games on Steam, Valve no longer needed to ship physical software boxes to retailers. Steam allowed developers to distribute their games instantly at no marginal cost on the platform, which particularly gave small developers access to the market so they could grow their own customer base. The cost savings for both game producers and players were considerable, slashing manufacturing and distribution costs of game discs significantly.

With more and more games available on Steam, customers grew as well, including those who hadn't played Valve's own games before. In 2021, Steam saw 2.6 million first-time buyers each month.[6] Valve captured new value by taking a cut on game sales that independent developers distributed on Steam, similar to Apple on its App Store. Steam had nearly thirty thousand games available on it as of May 2022, enabling millions of gamers to play online concurrently.[7] The commission from sales by other developers on the platform has well exceeded the revenue from Valve's proprietary software.

Being willing to open yourself to your rivals

Platforms involve connecting third parties. But some companies and organizations go a step further by embracing their competitors, too.

Our home school, HBS, is known for its case teaching methods, and Harvard Business Publishing Education (HBP Education) has long been a distributor of all the cases written by HBS faculty. As a not-for-profit division of HBP, it carries a catalogue of over thirty thousand cases, but only about 43 percent of them are from HBS faculty.

Over the years, HBP Education has grown significantly by working with more than fifty content partners, including other schools at Harvard and business schools around the world. Some of them are our competitor schools, such as Stanford Graduate School of Business or Columbia Business School. Although learning materials by faculty in other schools can potentially compete against those by HBS professors, the HBP platform is able to offer a wide variety of choices for users to purchase, growing the popularity of the platform substantially over the years.

When a firm opens itself up to rivals, it is possible that the rivals will have reservations. They may resist supporting the firm's platform because their participation might strengthen a competitor and they do not want to depend on a competitor to gain access to users. This is particularly true when the rivals are large. In such cases, to grow a platform, a firm may need to incentivize cooperation by sharing ownership of the platform.

Take the US banks as an example. They couldn't sit still and watch the increasing threat of PayPal/Venmo, Apple Pay, and Square Cash moving into retail payments. In 2017, the US banking industry rolled out Zelle, a person-to-person payment network in response to the popularity of digital payment.[8] Zelle was a joint venture by several large US banks. Its operator and controller, Early Warning Services, was a fintech company owned by seven major US banks, including rivals Bank of America and JPMorgan Chase, with an extensive list of lenders that support the networks.[9] Despite being longtime competitors, they are now sitting at the same table (platform) to face their digital competitors.

Zelle enables users to send money quickly from a banking app to friends and family, a much simpler method than traditional bank wiring services. Whereas its closest brand name competitor, Venmo, charges fees for several premium offerings, the Zelle app is free to use.[10]

While PayPal/Venmo still dominates the peer-to-peer payment market in the United States, Zelle saw a record growth with $806 billion in transactions in 2023.[11] Zelle doesn't have many of the same premium features offered by Venmo-like apps, but its position as a bank-based platform does offer increased customer trust—customers may view Zelle as a more secure payment option, especially for larger transactions, such as paying rent.[12] More importantly, its strong ties to different banks also helped Zelle tap into small and medium businesses as they seek safe and fast alternatives to sending checks. In 2023, payments received by small businesses via Zelle increased 44 percent from a year earlier, a strong force behind the platform's growth.[13]

By supporting a common platform, the banks were able to better defend themselves against fintech startups and tech giants such as PayPal/Venmo and Apple Pay.

Actively seeking to create value for other companies

Being willing to open up to third parties is the first step toward a platform strategy. Next, companies also need to identify platform opportunities in their businesses. One effective approach is to ask how a company could leverage its current resources and capabilities to create value for other companies, in addition to itself.

Telepass. Telepass, a company for which our colleague Chiara Farronato has conducted case research, initially handled electronic tolls on highways in Italy, France, Spain, and Portugal, later evolving into a mobility platform.[14] In 1990, it introduced an electronic toll collection (ETC) device that users place on their vehicle's dashboard. This device allowed them to drive through toll plazas without stopping, significantly improving the traffic flow between Milan and Turin during the FIFA World Cup. The plaza receiver would automatically deduct the appropriate fee from the user's account. By 2016, this business model had generated reliable annual revenue across Europe for the company.[15]

In due course, Telepass expanded its services, enabling users to pay for parking using its device. In the early 2010s, car and bicycle sharing services began to boom. To cater to consumers' new mobile needs, Telepass launched Telepass Pay, a digital payment mobile app, in 2017. Since the company already had the existing Telepass subscribers' bank account information, the addition of Telepass Pay was seamless. Users simply downloaded the Telepass Pay app, logged in using their existing Telepass credentials, and started using the new service instantly. Telepass Pay can be used to buy train tickets, rent scooters, and fuel up.

In 2019, Telepass began to act as a broker for insurance products, providing third-party insurance renewals via Telepass Broker to its customers. With the subscribers' permission, Telepass Broker can track when their policies are due for renewal and suggest a new offer.

With user consent, the data amassed by Telepass can potentially deliver more value to both insurance companies and users. Tolling data (with over 700 million toll transactions processed per year) and Telepass Pay data can help the company estimate users' risk profiles. For instance, the company has information about the distance between any two toll plazas and the time stamps of toll collections for each driver. Consequently, Telepass can calculate the average speed of a car. An abundance of data points on each user's driving speed can assist in determining a driver's risk profile. This information is valuable for creating customized insurance policies for each driver.

McGraw-Hill Education. McGraw-Hill Education, the over 130-year-old US educational publisher, also discovered new value creation opportunities during its twenty-year transformation from print to digital. In the pre-digital age, McGraw-Hill, as a publisher of textbooks in printed form, could not interact with readers directly because readers bought books from bookstores. In 2009, the company launched an all-digital teaching and learning platform for higher education and later expanded to the K–12 field. By acquiring software companies and applying AI solutions, McGraw-Hill attempted to turn the page from people reading books to books reading people. Based on their past reading and test-taking behaviors, the McGraw-Hill platform can point students to the right content in the book and curate personalized content.

About 75 percent of McGraw-Hill's higher education businesses, the biggest part of its portfolio, was already digital at the end of 2019; the pandemic accelerated the share to 89 percent in 2020.[16] The learning data that the software gathers also provides value and insights to teachers, helping them understand their students better.[17]

In 2020, McGraw-Hill took digital learning further by partnering with TutorMe (now called Pear Deck Learning) to offer on-demand tutoring for college students. After identifying areas that they struggled with, the students using McGraw-Hill's digital courseware are able to access a tutor to help them.[18]

Even if your firm cannot identify any opportunities to become platforms today, you should make attempts from time to time. As traditional businesses become increasingly digitalized, just like Telepass and McGraw-Hill, you will inevitably gain new resources and capabilities in this process. You may find that some of these resources or capabilities can be valuable to other firms, thus providing opportunities for the product-to-platform transformation.

Connecting your consumers

We have discussed how traditional businesses can transform themselves into platforms by sharing their resources and forging fresh connections with third parties. However, this isn't the sole approach. Traditional businesses can also discover platform opportunities by nurturing relationships among their existing customers and facilitating their interactions.

The Sephora Beauty Insider Community, discussed in chapter 1, is a prime example of this strategy. Sephora cultivated a community of its customers and empowered them to ask questions, share personal styles, exchange advice, and try out new products. Sephora facilitates the interactions among content seekers and content producers. Without a doubt, this strategy enhances the loyalty of Sephora customers, boosts their satisfaction with the brand's products, and subsequently, increases Sephora's sales.

Similar strategies have been used by other companies. Athletic products stores, such as REI and Dick's Sporting Goods, sell equipment, apparel, gear, and so forth to people with a variety of interests like golfing, hiking, swimming, cycling, and team sports. These stores also host events to bring customers together, sponsor local sports teams and leagues, hold clinics or classes, and encourage a social community of customers. This strategy emerged as a lifeline for several independent bookstores, enabling them to withstand competition from Amazon. This comes against the backdrop of thousands of independent bookstores closing down, unable to match Amazon's pricing, convenience, selection, and affordable, rapid shipping. While almost all these bookstores started digital channels, their experiences show that simply imitating Amazon and having an online presence is insufficient for surviving.

Powell's Books. Powell's Books in Portland, Oregon, managed to not just survive but thrive by focusing on providing engaging customer experiences, connecting their customers to build local community, and maximizing the advantages of physical retail.

Founded in 1971 in a former car dealership on a then-derelict corner, Powell's has grown into a Portland landmark and one of the largest new and used bookstores in the world. Its flagship location occupies an entire city block and contains over one million books.

In response to the rise of Amazon, besides building its digital capabilities, the bookstore focused on connecting readers and building community in several ways. First, Powell's sponsors several themed book clubs that meet regularly in its three stores. Readers bond over a shared joy of reading and discussion of books. Powell's provides meeting spaces and helps promote the clubs.

Second, Powell's hosts over five hundred author events each year, bringing readers together to meet and learn from their favorite authors in person. Readers can connect over shared interests in the author or genre. Powell's also frequently partners with libraries, schools, and nonprofits on special events, fundraisers, and educational programs. These partnerships bring various groups of readers together around a common cause or interest.

Powell's stores also offer ample comfortable seating, with couches and reading nooks throughout the store. Readers often strike up impromptu conversations with others about what they're reading or recommend books to each other.

Powell's bookstores are open daily until 9 p.m. The extended hours make it not just a shopping destination but also a place for connection—readers study together, meet up with friends, or just curl up with a book into the night.

All of these factors have turned Powell's into a platform that fosters real-world interactions and relationships among readers. Readers feel a sense of shared identity through their love of Powell's. Shopping at Powell's and attending their events has become a way for Portlanders to support local culture and connect with one another.

This brand positioning as an independent, knowledge-centered, and community-driven bookstore contrasts with Amazon's more transactional experience. Technology used to be seen as a threat, but Powell's has leveraged its website, social media presence, podcast, and e-newsletter to enhance in-store experiences and build community rather than replace them.

Home Depot. In addition to building communities among your customers, you can look for opportunities to facilitate transactions among your customers. You are more likely to find such opportunities if you serve different types of customers. Take Home Depot, a large home improvement retailer, as an example. Home Depot serves a diverse group of customers, including do-it-yourself (DIY) customers, individual home-owners who purchase products for home improvement, repair, and maintenance projects they plan to do themselves; professional customers (contractors, remodelers, repairmen, small business owners, and other professionals in the construction and home improvement industry); and do-it-for-me (DIFM) customers, who buy the products but hire professional installers for services like installation of carpets, cabinets, or large appliances. Home Depot provides installation services in many areas including flooring, cabinets, and windows by licensed and insured contractors who work with Home Depot.

To better serve its customers' needs, Home Depot introduced Pro Referral as its online platform after its 2012 acquisition of Redbeacon, a startup that matches homeowners with local home service providers. When a customer makes a purchase from Home Depot for a service requiring professional skills, they can describe the work they need to complete. Pro Referral then uses a proprietary algorithm to match the request with qualified local independent contractors.

This service adds value for both the DIY customers, who may decide a project is beyond their skill level, and the DIFM customers, who are looking for a trustworthy professional to install products. It also benefits its professional customers, who gain access to Home Depot's large customer base. Unlike many other digital platforms that match homeowners with service providers, Home Depot does not charge consumers or contractors for requesting or receiving referrals.

Didi Hitch. It is important to note that you need to create real value for your consumers when connecting them. Otherwise, the connection can be detrimental if it is misaligned with what your customers value about your product or service.

Didi Hitch, the carpooling service provided by Didi Chuxing (Didi) of China, one of the largest ridesharing companies in the world, provides a cautionary tale. For a while, Didi believed that users of its carpooling service would value the opportunity to interact with each other and make new connections with fellow passengers in the shared ride. To this end, once passengers were matched and confirmed for a shared ride on Didi Hitch, the company made passengers' profiles visible to each other within the app, including their names and profile pictures. Didi hoped that this feature would allow them to get a basic understanding of who they would be sharing the ride with and potentially establish common ground for conversation during the journey. They could also use an in-app messaging feature to discuss details, such as meet-up points and other arrangements related to the journey. Didi Hitch also embedded a group chat function within the app, enabling passengers to communicate with all the fellow passengers who were part of the shared ride.

However, most Didi passengers did not feel the need to interact with others in the car. Worse, such social features weakened privacy and safety protection of the passengers, which most passengers cared about. After several incidents of assault and harassment during the Didi Hitch rides that led to a public outcry, Didi suspended and later removed the social features and implemented stricter safety measures. Its focus shifted toward enhancing passenger safety and reestablishing trust.

When firms try to connect their consumers, it is thus imperative for them to conduct market research to identify the benefits as well as potential caveats from such connections, including how such connections may affect how consumers interact with firms' current offerings.

Managing operating model shifts

The transition to a platform model not only alters a company's business model—how it creates and captures value—but it can also fundamentally transform its operating model and organizational culture. Effectively managing this shift is crucial for ensuring the success of such a transformation.

Product- or service-based businesses focus on developing the best or most unique offerings, maximizing profits, and optimizing unit sales. These businesses usually operate within closed, proprietary systems and maintain full control over their operations. In contrast, platform-based businesses thrive by establishing a network of partners and users, promoting platform adoption, maximizing interactions and transactions, and managing and influencing partners and users. They operate within open or shared systems. As the value of a platform stems primarily from its user network, it increases with the number of users.

The success of a platform-based business thus relies on its ability to attract a critical mass of users and foster engagement, making it more attractive over time. The platform can then more easily capture value with a large, loyal user base. Consequently, firms generally need to establish different key performance indicators (KPIs) for their platform businesses in the early stages. Instead of focusing on profits, firms should emphasize user adoption and engagement. Once the platform business gains traction, it can then concentrate on value capture opportunities. In the early days of Zé Delivery, for example, while employees at its parent firm, Ambev, earned variable compensation based on whether they met sales targets, Zé's performance indicators were largely based on client satisfaction, such as net promoter score (NPS), number of active app users, and purchase frequency. A common mistake many organizations make is applying the same KPIs of their traditional business model to their platform business too early.

Second, companies must not underestimate the new capabilities required for their platform businesses. For example, Nigeria-based EbonyLife Media discovered that adopting an on-demand streaming platform model wasn't necessarily the best solution for its growth. Established in 2012 to share high-quality African stories globally, EbonyLife began by producing content for its pan-African linear television channel and later expanded into movie production. It had great success in creating blockbuster and critically acclaimed content viewed by an increasingly global audience. Despite its success in producing content, it faced difficult decisions on how that content should be distributed, especially beyond Africa.[19]

It launched an on-demand platform, EL ON, in 2014. The company struggled to find suitable local software developers and ultimately hired a French team to develop the necessary software solutions. To retain subscribers, EbonyLife needed a constant supply of fresh content, which it lacked the funding to produce or source. Nigeria's unreliable internet connectivity also negatively impacted the streaming experience, causing some users to unsubscribe. Becoming an EL ON subscriber in Nigeria often involved a manual process because most Nigerian customers relied on bank transfers for payment. Consequently, EbonyLife could not provide them with immediate or automatic access upon payment.

EbonyLife attracted a mere eight thousand new subscribers between 2018 and 2019, falling far short of its one-million-subscriber goal. CEO Mo Abudu began doubting the viability of transitioning EbonyLife to a streaming platform business model. Abudu then explored coproducing content with Netflix, Sony, and other global partners for distribution on their platforms. These international streaming channels provided upfront financing, reducing EbonyLife's production-related financial risks. By 2021, EbonyLife had phased out EL ON to focus on content production.

Third, traditional businesses need to overcome the challenges in finding the right talent to drive the new platform business. In the case of Telepass, it sought out new hires skilled in engineering and data science. Telepass also entered into a three-year partnership with a system integration company called NTT Data in 2018 to oversee the transition to the cloud. Telepass planned to hire some NTT Data employees at the end of the partnership to ensure that it could continue to manage the new systems on its own.[20] This approach allowed Telepass to leverage the knowledge that they had outsourced early on and bring it in-house.

When firms start to facilitate new transactions with their platform model, they often need to equip themselves with new industry-specific knowledge. With the launch of Telepass Pay, the company began recruiting people from telecommunications and from banking, as well as from tech giants, to complement their expertise in transportation. Many new hires were also young, with the average age of Telepass C-suite executives being just over forty years old as of 2021.[21]

Finally, when traditional businesses incorporate platform business models, they often create separate teams or divisions for their traditional and platform operations, each with its own set of goals and KPIs. This structure enables the specialization of skills, resources, and processes tailored to each business model. However, cultural clashes can still arise when a company attempts to foster cooperation between the two models to capitalize on the combined strengths of both. Managing these distinct cultures can pose a significant challenge for any organization.

Ambev's experience provides a useful example. Since Zé's inception, its founders and Ambev intentionally set out to create a distinct culture for the startup. Ambev hired external consultants and studied the cultures of successful disruptive firms like Netflix, Google, and Amazon in order to establish what a company of the future should look like, infusing Zé with traits from those benchmark companies. Its culture revolved around customer centricity and a long-term business focus, which differed from Ambev's short-term, quarterly profit focus.

Zé employees deliberately adopted different approaches to distance themselves from the traditional business. For instance, while Ambev used Microsoft Azure, Zé engineers opted for Google Cloud. Simultaneously, employees in the traditional business resented the new hires at Zé, believing that they made all the money while the young engineers squandered it. This dynamic created animosity and tension within the company. Cultural differences and divergent profitability perspectives became increasingly evident as Zé grew larger.

Zé's then-CEO Rodolfo Chung devoted considerable effort to managing these cultural differences, promoting respect for both cultures, and encouraging communication and cooperation. He compared the relationship between Ambev and Zé to that of parents and children. In this analogy, children should be grateful for their parents' support and investment in their education, recognizing that they would not be where they are today without it. As a result, children should show respect to their parents. They can either refuse to engage with their parents due to generational differences, or they can bravely initiate conversations to share differing perspectives and foster mutual understanding. After all, children can often learn valuable lessons from their parents' experiences. Similarly, parents should strive to understand their children better, recognizing that it is in the family's best interest to support their children's success and grow with them.

As these examples illustrate, to successfully navigate the transformation to a platform business model and unlock new opportunities for growth, traditional businesses need to recognize these operational model shifts and actively respond to them.

Things to Look Out For

It's worth noting that a platform business model isn't always the best choice for an organization, even if it possesses the necessary capabilities for the transformation. One significant drawback of a platform model, compared to a product or service model, is that firms may need to relinquish some control, as they now facilitate transactions between consumers and third parties instead of providing these complementary products or services directly.

Two examples from the US grocery delivery sector, Instacart and Weee!, illustrate different approaches. Both companies gained popularity during the Covid-19 pandemic. Instacart, founded in 2012, one of the largest and most widely available grocery delivery services across the United States, started off with a light-asset platform model. Instead of building warehouses and physical distribution centers, Instacart partnered with existing grocery stores and contracted personal shoppers. The more than three hundred partner stores of the delivery platform range from most popular national and regional supermarkets, including membership stores such as Costco and BJ's Wholesale, to local specialty shops, and even pharmacies.

Instacart doesn't own any of the stores nor does it employ the shoppers permanently. It creates and captures value from the platform it established that links customers, shoppers, and groceries together.

Unlike the online grocery giant Instacart, Fremont, California–based startup Weee! began with the direct sales model, specializing in selling foods for Asian and Hispanic cuisines. To ensure that product prices on Weee! are competitive to store prices, the company managed its supply chains directly. It built its own warehouses and worked with part-time drivers. As of May 2022, orders are packed and shipped from Weee!'s eight fulfillment centers across the United States.[22]

There's no one-size-fits-all solution to choosing between a platform model that relies on third parties and a direct sales model. Each has its advantages and drawbacks. In the platform model, partnering with established stores (the Instacart model), gives the company an option to offer a variety of products without building its own warehouses. However, in this model the company loses direct quality control. In addition, Instacart and grocery stores both need to capture value, making product prices higher and less appealing to price-sensitive customers. In contrast, the direct sale model provides Weee! both great quality and cost control but limits its product selection and adds operational burden.

Interestingly, both companies are shifting toward a hybrid model. Instacart plans to operate its own network of warehouses, which it calls nano-fulfillment centers, a move that would help it to store products and launch a fifteen-minute delivery service.[23] Weee! has also added many third-party sellers to offer more product variety.

The hybrid model mitigates the limitations inherent in each individual model. However, it's crucial for companies to recognize the capabilities they must develop to successfully implement it. Therefore, adopting a hybrid model should be contemplated only after a company has gained the capabilities to support one model and is ready to take on a different one. It is also important for companies to be mindful of the potential cannibalization effects between the two models, which may affect the motivations of third parties. (We will revisit the competition between third-party products and those offered by the platform in chapter 5.)

Digitalization has increased connectivity between products or services and their users, allowing these offerings to serve as channels for connecting users to new products and services.

Any company today should therefore actively consider product-to-platform opportunities.

This chapter outlines fundamental principles for identifying potential platforms within businesses and facilitating their transition. Begin by considering how you can create value for other companies, including competitors. Then carefully contemplate the transformation necessary to achieve success.

CHAPTER 4

Grow Your Own Ecosystem

S hein, the Chinese ultrafast-fashion retailer, tops the list of the world's most mysterious, fastest-growing companies. Pronounced "she-in," the little-known name before the Covid-19 pandemic has grown and grabbed the title of the fashion world's most Googled brand, surpassing Zara and Nike.[1]

Founded in China in 2008, it evolved from a small retailer to a cross-border retail platform and branded Shein for women's apparel.[2] Known for its ultracheap and trendy clothing, the company designed and sold apparel, beauty, and lifestyle products exclusively abroad in more than 150 countries as of 2024.

Despite the criticisms that its business model encourages wasteful consumption and other concerns such as its supply chain practices, potential copyright infringement, and lack of data protection, Shein was reported to have a $65 billion valuation in 2023 and became the world's most valuable startup after ByteDance, SpaceX, and Ant Group.[3]

Shein's growth in the digital age owes to the ecosystem it has cultivated over the years. A company's ecosystem includes all entities and individuals that are affected by, and can affect, the company's business. Just as in a biological system, where a balanced ecosystem benefits all its members, the same principle applies in the business world. This chapter focuses on how traditional businesses can successfully orchestrate their own ecosystems.

Shein's Ecosystem

Shein's ecosystem involves many participants. First, Shein leveraged many clothing manufacturers in China. It operates a digital supply chain that brings together about six thousand low-cost suppliers in southern China, home to thousands of garment workshops and dealers.[4]

The center of Shein's supply chain operations is in China's southern business hub of Guangzhou, which is close to the country's warehouse centers in Foshan and an international airport in the region. Shein benefited from the supply chain capability and infrastructure in Guangzhou, which were built more than thirty years ago and evolved together with China's textile export and e-commerce industries. Shein deliberately picked those small and medium-sized factories because they possessed less bargaining power compared to larger factories, and they were generally more willing to accept small batch orders. Despite their size, many of these seasoned factories have previous experience working with e-commerce giant Alibaba. They are often situated within a five-hour drive from Shein's Guangzhou hub, thereby accelerating communication and collaboration.[5] Shein set stringent requirements for delivery timelines and aggregated the unused production capabilities of its partner factories.

The company has invested in its manufacturing execution system (MES) since 2016, which enables real-time supply chain management. Shein suppliers are required to use the MES for seamless communication during the manufacturing process. Shein's MES provides accurate and timely feedback and is highly regarded for its user-friendliness.

By 2022, Shein launched six thousand new products daily, the prices of the products averaging $6 to $25 per item.[6] Shein's in-house analytics tools allowed manufacturers along its supply chain to receive real-time data on which items were trending and how well certain items sold.[7] Items that sold well were automatically reordered, while those with poor sales were phased out. The turnaround time for each item was as short as twenty-five days, while it would traditionally take months for many retailers.[8] Supported by the data, the company would ask its global designers to work on the design for its Chinese suppliers to produce.

Besides these clothing manufacturers in China, a growing number of consumers in Europe, the United States, or around the world participated in Shein's ecosystem. The company created value for them in many ways. In addition to its low-cost supply chain, as a cross-border e-commerce company, Shein took advantage of favorable tariff policies for parcels with low values, which allowed it to further lower its prices by 15–20 percent to attract overseas customers.[9]

To accelerate the delivery time, Shein heavily invested in warehouses and distribution centers globally. Its distribution center in Whitestown, Indiana, could reduce shipping times by up to four days, while Shein planned to open facilities in Southern California and the Northeast United States.[10] In April 2023, it announced plans to localize production in Brazil by partnering with two thousand textile factories in the country within five years, as well as in India and Turkey.[11]

Cross-border logistics companies are also participants in Shein's ecosystem. It offers free standard shipping, which averages about eight to ten days for delivery and free return services when a customer spends a minimum amount. A Shein order is usually shipped from its warehouses in southern China, put on a cargo plane (for example, it partners with China Southern Airlines Logistics Company) and delivered directly to the customers abroad.[12] Its delivery partners in the United States include DHL and YunExpress, and Yida-Ex is its partner in the United Kingdom.[13]

Many perceive Shein as a threat to Amazon. But Shein also incorporated Amazon into its ecosystem by listing thousands of its offerings on Amazon's marketplace, including some of its bestsellers.[14] In the meantime, Shein also opened itself up to third-party vendors to become a marketplace.[15]

TikTok became an important participant in Shein's ecosystem as well. Shein started early on its digital marketing on TikTok by paying endorsements and brand hashtags to social media influencers to promote its brand.[16] Shein became the most talked-about brand on TikTok in 2020.[17] The massive content generated on TikTok effectively helped Shein grow its main customer base: female Gen Z consumers and the low-to-middle-income population in the United States and Europe.[18] These consumers are price conscious, but selling to them still yields higher margins compared to domestic buyers in China, who exhibit a higher level of price sensitivity. (See figure 4-1 for an illustration of the main players in Shein's ecosystem.)

Every company has its own ecosystem, but few manage to leverage and expand it for growth as effectively as Shein. The Shein example also illustrates the difference between an ecosystem and a platform. Though the terms "ecosystem" and "platform" are often used interchangeably, an ecosystem is a broader concept encompassing a network of interconnected businesses, organizations, and individuals that cocreate and exchange value. An ecosystem can sometimes

FIGURE 4-1

Main players in Shein's ecosystem

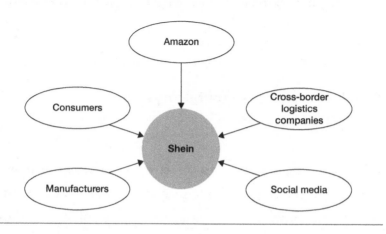

encompass a platform firm but incorporate many other elements such as suppliers, distributors, competitors, and government agencies. Shein's ecosystem, for instance, includes cloth manufacturers and logistics service providers such as postal services.

Recognizing that they may lack the requisite capabilities or resources to effectively compete against tech giants, many traditional businesses today eagerly focus on building and expanding their ecosystems. This chapter illustrates a few core principles that traditional businesses can use to successfully orchestrate their own ecosystems.

Principles for Growing Ecosystems

Traditional businesses, often accustomed to a more self-contained and linear approach, might find it challenging to adjust to the open, dynamic, and interdependent nature of ecosystem strategies. Hence, this transition necessitates a paradigm shift in strategic thinking.

We begin with the first shift—viewing tech giants as participants of your own ecosystems.

Flip the perspective: Your ecosystem, tech giants included

Often, traditional businesses perceive themselves as mere participants in the ecosystems of tech giants when collaborating with them. However, it's crucial to keep in mind that these tech giants are also working to create value for you in these relationships.

By shifting this perspective and considering these tech giants as participants within your own ecosystem, you can become more proactive in harnessing their resources, such as data and infrastructure, for innovation and growth. In today's digital age, it isn't daunting to utilize resources from tech giants to fuel innovation. Companies can tap into a plethora of available resources—be it the myriad Amazon product reviews, a rich collection of free videos on YouTube, or Google's open-source algorithms. Thanks to these tech giants, a significant portion of your ecosystem is already established and readily accessible.

Anker: Supercharging its ecosystem. Anker Innovations is a company that emerged from Amazon and has been leveraging the tech giant's ecosystem since its inception. You'll likely come across the Anker brand if you ever search for a power bank or an iPhone cable on Amazon.

Anker was one of Amazon's native brands. Founded in 2011 by ex-Googler Steven Yang, Anker sells digital accessories—batteries, cables, chargers, and so forth. These products are also offered by hundreds of other manufacturers or agents. In addition to renowned branded items like Apple's cables, Amazon sells a wide array of private-labeled digital devices at competitive prices.

Nevertheless, despite facing competition from these tech giants and other manufacturers, Anker's growth continued unabated. Over the span of a decade, Anker emerged as the most popular brand of portable battery power banks on Amazon.[19] Its products are known for their good quality at a fair price. Anker, which went public in 2020, had a market cap of 30 billion yuan ($4.2 billion) as of January 2024.[20]

Yang credited Amazon for the company's success. In addition to Amazon's established infrastructure that helped it save on retail and logistics costs, "the key to building high quality and innovative products is to listen to the customers. Amazon reviews are actually the single most important input to our new product development process. We make sure that our new products start from the needs that customers express," Yang said in an interview in 2016.[21]

Anker, focusing on the Amazon channel, has developed an efficient system to capture and promptly address customer feedback, known as the voice of the customer (VoC). For instance, when consumers reported that data cables were prone to damage after some use, engineers transformed this issue into a clear, quantifiable quality standard: the cable should withstand ten thousand swings at 120 degrees under a 10-kilogram load without breaking.[22]

Beyond the online VOC feedback system, the company also discerns user pain points through methods such as focus group research and expert interviews. For example, Anker discovered that the apparent lack of interest among female users in charging accessories was due to the absence of products tailored to their needs. To bridge this gap, Anker ingeniously created a compact, high-speed power bank with the appearance of a lipstick, offering a small, lightweight solution that is perfect for carrying in a purse. With a great understanding of customer needs, together with heavy investment in product research and engineering—research and development (R&D) workforces account for about 47 percent of Anker's more than four thousand employees globally—Anker was able to develop many hit products, and its products won several top international awards.[23]

While more than half of Anker's 2022 sales of 14.25 billion yuan ($2 billion) still came from Amazon, the dependence has decreased from 80 percent in 2016.[24] Leveraging the momentum and great reviews on Amazon, the company started selling through its direct-to-consumer websites as well as penetrating other sales channels. For example, Anker also sells products in Apple's official direct-sale stores, selling similar products next to Apple's own products such as charging cables.[25] Anker has also managed to partner with offline retailers Best Buy, Costco, Target, and Walmart and sell through their channels. Through this strategy, Anker has managed to develop an ecosystem that encompasses several giants including Amazon and Apple, thereby further solidifying its position in the market.

Traditional businesses, much like Anker, should embrace the mindset to proactively view tech giants as potential participants in their ecosystems. Rather than merely utilizing these giants as channels to broaden their reach, traditional businesses should leverage these tech companies to boost their own capacity for innovation. It's this enhanced ability to innovate that not only provides traditional businesses with a competitive edge against tech giants but also draws multiple tech giants into their ecosystems. This, in turn, strengthens their bargaining position against each of these technology behemoths.

To hub or not to hub: Finding the right
path to grow your ecosystem

When building an ecosystem, companies have choices: they can be a hub or a non-hub. Both have pros and cons.

A hub in an ecosystem is a central organization that provides essential services, technology, or connections used by a large number of other participants. Orchestrating an ecosystem as the hub can be a significant undertaking. It often requires a firm to make up-front risky investments with the hope of reaping the rewards at a later date. For traditional businesses unaccustomed to acting as a hub and lacking the commitment to invest time, energy, and resources, undertaking such an endeavor could pose significant risks.

Anker's journey shows that to build an ecosystem you don't necessarily have to be a hub. Anker does not aim to become a hub for participants in its ecosystem nor is it essential to the survival of the participants in its ecosystem.

While it may seem disadvantageous for non-hub players to depend on other participants for various needs and not to possess the same power enjoyed by ecosystem hubs, firms can significantly reduce their downside risks and quickly begin to benefit from the ecosystem.

Ping An's ecosystems. Many companies with rich resources seek to become hubs when they build their ecosystems.

Ping An Group, China's largest insurer, took this approach and sought to become a hub in five ecosystems, including financial service, health care, auto services, real estate, and smart city ecosystems.[26] We talked briefly about its history and development in chapter 2: Ping An started as a thirteen-person small property and casualty insurance office in China's southern city of Shenzhen in 1988 and grew to become a conglomerate of 1.4 million sales agents at its peak in 2017.[27]

Over the years, Ping An implemented its finance plus ecosystem strategy through three steps.

First, it used technology to enable and increase the competitiveness of its core business of financial services. Second, it leveraged technology to enable five new ecosystems. When building each ecosystem, Ping An developed scenarios first, built traffic, generated revenue, and eventually hoped to make profit. After building the ecosystems, the company also hoped to use those ecosystems to grow its financial services.[28]

Ping An picked these five ecosystems because the business opportunity in each was big enough: They created many touch points for Ping An to better understand a customer's needs, and they could leverage Ping An's strength in financial services. (See figure 4-2 for Ping An's ecosystems.)

FIGURE 4-2

Ping An's ecosystem

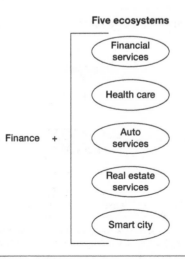

Ping An's experiences with growing different ecosystems varied, offering insights into the opportunities and challenges of becoming a hub.

Consider the auto services ecosystem. To establish its own ecosystem and become a hub in this area, Ping An acquired Autohome, China's leading automobile information portal and an online car-trading marketplace, in 2016.[29]

With an average of 46.9 million mobile daily active users then, Autohome offered a wide range of services and content that enabled users to research and purchase vehicles.[30] The portal also enables auto dealers to sell vehicles, with over 80 percent of its revenue coming from advertisements and fees from automakers and dealers.[31] Integrating Autohome with Ping An's existing services, including auto insurance, Ping An Bank, and Ping An Financial Leasing, the company aimed to support users throughout their entire car ownership journey— from purchase to sales, maintenance, and leasing—while providing a valuable channel for auto dealers to reach customers. (See figure 4-3 for Ping An's auto services ecosystem.)

FIGURE 4-3

Ping An's auto services ecosystem

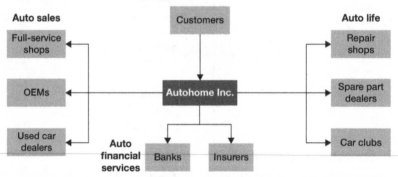

Source: Adapted from Feng Zhu, Anthony K. Woo, and Nancy Hua Dai, "Ping An: Pioneering the New Model of 'Technology-Driven Finance,'" Case 620-068 (Boston: Harvard Business School, 2020).

Ping An's health-care ecosystem expansion had more challenges. At its core was Ping An Good Doctor (PAGD), a telemedicine provider that Ping An launched in 2015, offering a comprehensive loop of services for patients, health-care providers, and payment solutions.

PAGD aimed to provide a family doctor for every household, an e-profile for every individual, and a health-care management plan for everyone, leveraging "mobile medical + AI technology."[32] Distinguishing itself from other online health-care providers in China, PAGD recruited over one thousand doctors from top hospitals to work full-time, aiming to enhance the quality and reliability of its telemedicine services.[33] PAGD demonstrated significant value creation and experienced rapid growth during the Covid-19 pandemic, reaching over 440 million registered users as of June 30, 2022.

However, since its inception, the company has incurred significant losses, including a $220.5 million loss in 2021.[34] Significant investment in its medical team and low acceptance of internet-based medical services in China are major contributing factors. In addition, while telemedicine providers appear to have access to many health-care resources, they ultimately depend on traditional, offline hospitals for delivering care. At the same time, the top-tier physical hospitals often lack the spare capacity or inclination to support telemedicine providers.

PAGD's competition with Alibaba Health Information Technology and JD Health International, which are the digital health arms of two deep-pocketed Chinese e-commerce giants respectively, further complicated PAGD's path to profitability.

While PAGD was still struggling, Ping An's other ecosystem, real estate services, could not proceed further. The company stopped pumping money and shut down its property-listing portal in 2019, five years after its launch.[35]

China's property industry highly relies on brokerage agents. Ping An had expected some of its 850,000 life insurance agents could help sell houses and apartments, as they had known their customers well.[36] But life insurance agents venturing into real estate sales face significant hurdles, necessitating a comprehensive understanding of property sales practices. Moreover, the real estate sector in China is known for its intense competition, with traditional brick-and-mortar brokerage models and conventional agents retaining their strength and resilience. Furthermore, these traditional brokerages are also actively digitalizing their operations.

Notably, China's largest real estate brokerage, Lianjia, set up its own digital property listing site, Beike, while newcomers Alibaba and Tencent also entered the crowded space through their portfolio companies.[37] To make the matter worse, the Chinese government's curbs on real estate financing took a toll on some of Ping An's high-yield property loan products.

In comparison, DBS Bank took a different approach to enter the property market. Instead of developing its own property marketplace, it partnered with EdgeProp and Averspace, two existing property marketplaces, to quickly increase the number of listings to around one hundred thousand from both agents and owners. It instead focuses on areas where DBS has expertise such as offering a home financial planner that helps first-time buyers determine their affordable price range based on their monthly cashflow. It also offers end-to-end paperless transactions, from check-free payments to digital documentation.[38]

The approach reduces the investment and its associated risks significantly.

The experiences of Anker, Ping An, and DBS highlight the strategic importance for companies to decide whether to position themselves as hubs or non-hubs within their respective ecosystems. As non-hubs, companies can swiftly utilize their existing capabilities to collaborate with others. To establish themselves as hubs, although the growth potential is greater, they often need to develop new capabilities to attract and retain participants.

Even for resource-rich firms like Ping An, developing an ecosystem as a hub in a competitive environment presents considerable challenges. Most traditional businesses lack the resources of a company like Ping An. Therefore, even if aspiring to become a hub, beginning as a non-hub player can be beneficial. This approach allows companies to empower others while gaining valuable experience, understanding customer and partner needs, and accumulating the skills and resources needed for future hub status.

For companies entering an ecosystem late, adopting an acquisition strategy can be an effective way to establish themselves as hubs. Ping An successfully utilized this tactic in its auto services ecosystem. In a similar vein, Adidas entered the digital fitness market—a field Nike had already ventured into— by acquiring the running app Runtastic in 2015, which then had seventy million users. This acquisition quickly became a cornerstone of Adidas's digital strategy. Today, over 170 million people use Adidas Running to track more than ninety sports and activities.[39]

Firm footing: Building your ecosystem on a solid foundation

Successful ecosystems often begin with solid and thriving products at their core. To ensure sustained growth of an ecosystem, a company must prioritize the defensibility of its product.

In the case of Anker, the company was able to convince industry leaders, such as Apple and Costco, to join its ecosystem because of its relentless efforts into identifying customer needs and R&D to introduce hit products.

Nike's ecosystem. In chapter 2, we explored how Nike orchestrated a fitness and wellness ecosystem centered on its physical products (like shoes and apparel) and digital offerings (such as the Nike+ Run Club and Nike Training Club apps). However, this journey was met with its own set of hurdles.

One of Nike's initial ventures into fitness was the 2012 launch of the FuelBand, a wrist-worn fitness tracker with social features. Its key feature, NikeFuel, aimed to offer a universal metric for tracking various activities. This was part of Nike's strategy to blend digital technology with its physical products, fostering a fitness and wellness ecosystem. Nike soon realized, however, that its market position in the FuelBand was not defensible. The wearable technology market rapidly evolved with competitors like Fitbit and, at the time, Jawbone UP, and the emergence of smartwatches, especially the Apple Watch, which offered broader functionalities. Acknowledging that hardware innovation was not its strength, Nike discontinued the FuelBand in 2014 to concentrate on software. Its digital products, Nike Run and Nike Training Club, which better aligned with its core strengths in branding, content creation, and community engagement, emerged as hubs of its fitness and wellness ecosystem.

Xiaomi's ecosystem. In China, Xiaomi also learned the importance of a defensible smartphone business to its internet of things (IoT) ecosystem from its experiences. Founded in 2010 as primarily a smartphone brand, Xiaomi has evolved to develop the world's largest IoT ecosystem. Since 2014, Xiaomi has invested in over five hundred IoT startups and partners, holding minority stakes.[40]

According to its founder, Jun Lei, Xiaomi's IoT strategy was devised to circumvent China's tech giants at that time (Baidu, Alibaba, and Tencent) who dominated the digital markets. Xiaomi initially offered mobile accessories like headphones and power banks, expanding in 2016 to include smart home products, such as rice cookers, televisions, and washing machines, and later adding lifestyle items. (Figure 4-4 shows Xiaomi's IoT ecosystem products.) Xiaomi's ecosystem evolved into AIoT (artificial intelligence of things), combining AI and IoT.

Xiaomi's smartphones became hubs to this ecosystem, designed to work seamlessly with Xiaomi IoT products. Through Xiaomi's MIU OS (now evolved to HyperOS), an Android-based operating system, and apps, users can manage a range of devices, further integrating the brand into their lives. Xiaomi collected extensive data from millions of connected devices, providing ecosystem partners with insights for data-driven decision-making and targeted marketing strategies.

Products in Xiaomi's IoT ecosystem

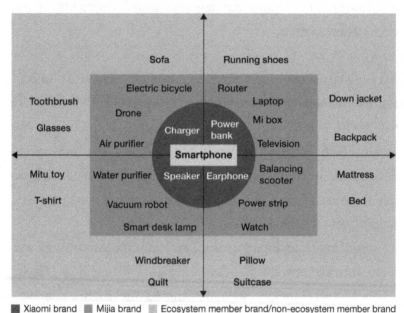

■ Xiaomi brand ■ Mijia brand ▓ Ecosystem member brand/non-ecosystem member brand

Note: Mijia is a brand in the Xiaomi ecosystem, which is also known as Xiaomi sub-brand.

Source: Adapted from "An In-depth Look at the Logic of Xiaomi's Ecosystem" [in Chinese], May 8, 2019, https://itw01.com/UA439ER.html.

Xiaomi's smartphone market success was partly due to its lean business model, supported by a mature supply chain and marketing strategies like hunger marketing (a method whereby Xiaomi purposefully produces fewer units than the demand, thereby creating a sense of exclusivity and fueling hype around their products). However, its weaker R&D foundation led to a slowdown in its growth momentum, especially as competitors quickly introduced new models. Meanwhile, the global smartphone market is experiencing a weaker growth rate, as consumers are waiting longer to upgrade to the new models.

Facing setbacks in the smartphone market, Xiaomi seemingly pinned its hopes on its ecosystem for business revival. The keyword *ecosystem* was mentioned 122 times in its prospectus when the company went public in 2018.[41]

Yet, an ecosystem-focused strategy did not resolve the key growth impediment—its smartphone business. Although the saturated smartphone market may seem unattractive, the frequent use of smartphones by consumers can boost sales of ecosystem products and foster loyalty among ecosystem partners, thereby discouraging them from forming affiliations with competing ecosystems. If the company's core product of smartphone sales continued to slide further, Xiaomi's ecosystem could collapse.

Recognizing the pivotal role of its smartphone business, Xiaomi, in 2020, repositioned its strategy to focus on smartphones. Founder Lei made it clear that smartphones would remain a cornerstone of its future business model, and AIoT would build on this core business.[42] This renewed focus and increased R&D, especially in high-end smartphones, led to a surge in shipments in 2021, securing its place as the third-largest player in the global smartphone industry.[43] The growth of Xiaomi's ecosystem will critically depend on its ability to maintain the popularity of its smartphone business or to build and grow a new defensible ecosystem core (e.g., cars).

Both Nike's and Xiaomi's experiences underscore the importance of a defensible core product or service as the foundation of a successful ecosystem.

Understanding and managing co-innovation risks in your ecosystem

Some firms focus on harnessing existing resources within their ecosystems, much like how Anker leveraged Amazon,

Costco, or Apple to reach their customer base. In these cases, the resources already exist within the ecosystem; firms merely need to identify how to utilize them effectively. Quite often, companies strive to collaborate with ecosystem partners to codevelop new products or services. In these instances, they face co-innovation risks because the ultimate success of their innovation relies on the joint efforts and collaboration of all parties involved.[44]

KT and 5G. Telecom companies face a challenge: they provide vital network infrastructure and handle extensive data, yet much of the economic value is captured by over-the-top (OTT) content providers like YouTube, Meta, and Netflix, which deliver media directly over the internet.

In 2016, amid slow domestic growth, South Korea's second-largest wireless carrier and a leader in network technologies, KT Corporation, pivoted to exploit its strengths in infrastructure and network by spearheading the industry's shift to 5G services.[45]

KT's 5G initiative faced co-innovation risks, requiring not just its own technology development but also the need to secure 5G handsets and compelling digital content and applications that would take full advantage of 5G technology.

The success of such co-innovation depends on the probability of each partner meeting their commitments. For instance, if KT's 5G technology readiness, the availability of 5G handsets, and the emergence of a compelling 5G killer app each had a 70 percent probability, KT's overall chance of success would only be 34 percent ($0.7 \times 0.7 \times 0.7$) when treating these occurrences as independent events.

KT utilized several strategies to alleviate co-innovation risks. First, the company fostered collaborations with competitors to set industry guidelines for 5G equipment.[46]

Second, KT encouraged confidence in its 5G initiatives by announcing its strategic direction and timeline to the world in 2015 and showcasing its 5G prototypes during the 2018 Winter Olympics in Korea. This behavior contrasted with the more secretive approach of many companies, including Apple, which usually keep innovations confidential until an official product release. By sharing the progress, KT aimed to inspire other industry players to embrace 5G innovation.

KT also focused on developing killer apps, like ultra-high-definition (UHD) content and exploring VR and AR technologies, despite uncertainties about their demand on small screens. Additionally, KT collaborated with Hyundai in the autonomous vehicle sector, recognizing the transformative potential of 5G in real-time communication for transportation, though this sector itself requires many co-innovations to take off.

In April 2019, KT launched the world's first commercial 5G network, positioning South Korea as a leader in 5G commercialization.

Yet, four years later, the government's assessment revealed that the country's 5G progress fell significantly short of initial expectations.[47] The promised introduction of new AR, VR, 3D content, and IoT services did not materialize. The high frequency of 5G signals, which limits penetration and requires dense base station networks, along with the uncertainties in value capture, deterred aggressive infrastructure investment by KT and its competitors.

KT's 5G journey highlights the challenges of managing co-innovation risks when firms leverage their ecosystems to innovate. Companies thus should try to build ecosystems that require fewer risky co-innovations."

Consider Ping An's latest ecosystem initiative as an example. In 2022, Ping An launched home-based elderly care services to pursue its finance + elderly care ecosystem.[48] With a societal shift toward separate living for the elderly and their children and a preference against nursing homes in China, home-based care is increasingly appealing.[49]

The market, still nascent in China, offers Ping An the opportunity to introduce low-cost, mass-market elderly care, utilizing its experience from other ecosystems. For example, it can employ PAGD's medical team for online diagnoses and treatments, and to facilitate fast-track in-person care at partnered hospitals. Ping An also integrates its technology and AI expertise, using health monitoring devices for remote monitoring of vitals like blood pressure and heart rate, and detecting risks such as falls, sleep disturbances, and gas leaks.

By leveraging many in-house capabilities, the company can reduce co-innovation risks. True, Ping An will still need to rely on a number of partners, like hospitals and health monitoring device manufacturers, to build this ecosystem, but most of these resources already exist and do not pose significant co-innovation risks. Despite uncertainties about Chinese elders' acceptance of this technology-driven care approach, the limited co-innovation risks allow the company to focus on addressing its own execution risks, rather than those tied to its partners.

As these stories demonstrate, co-innovation within an ecosystem necessitates looking beyond one's capabilities to consider risks associated with interdependency within the entire ecosystem and devising proactive strategies to minimize these risks. Generally, traditional businesses will find more success when they largely leverage existing resources in their ecosystems to drive new value creation, rather than depending on many uncertain innovations by their ecosystem partners.

Things to Look Out For

Ecosystem strategies have undeniably become an essential path for the evolution and growth of traditional businesses. They offer opportunities to leverage shared resources, boost innovation, and secure a unique competitive edge through managing a network of customers, suppliers, competitors, and partners. Yet, designing and implementing these strategies is not without significant challenges.

This chapter emphasizes the need for the right mindset and flexibility in deciding whether to become a hub. It also highlights the importance of building a robust ecosystem foundation and effectively managing co-innovation.

As our examples show, businesses must develop their own tailored ecosystem strategies rather than mimic tech giants. This requires a deep understanding of the company's strategic objectives, resources, and capabilities. Only with this understanding can a business effectively identify and collaborate with partners who will truly complement and strengthen its operations, thus achieving greater success with a mutually beneficial ecosystem.

CHAPTER 5

Manage Frenemies

To be, or not to be," pondered Hamlet over four hundred years ago. Today, many traditional retailers find themselves grappling with a comparable question: "To sell, or not to sell (on Amazon)."

X Fire Paintball & Airsoft (X Fire), a retailer of paintball guns and supplies based in Auburn, Massachusetts, for over twenty-five years, saw a significant surge in sales once it began selling on Amazon in 2012. Online sales constituted 80 percent of the company's total sales. Steve Herbert Sr. and Steve Herbert Jr., the father-and-son ownership team, found that the benefits greatly outweighed the fees that X Fire had to pay Amazon.[1] Like a thousand other paintball dealers, Amazon has helped the company expand their customer base tremendously. In the words of the two Herberts, their business would have been dead without Amazon.

However, just four years later, sales of X Fire's top-selling products nose-dived. The Herberts swiftly uncovered the cause: Amazon had begun sourcing the same products directly from their suppliers, essentially competing with them head-to-head: a friend had turned into a foe.

Like X Fire, numerous enterprises, primarily small brick-and-mortar stores, must navigate intricate relationships with tech giants. Most traditional businesses neither aspire to nor possess the resources to become tech behemoths themselves. Given the market power of these tech titans, smaller players often find themselves obliged to collaborate with these giants.

As in the case of X Fire, the partnership with tech giants generates value and opens up opportunities. It allows them to access a vast customer base that was previously beyond their reach prior to the advent of the tech giants. During the Covid-19 pandemic, many brick-and-mortar restaurants were barely surviving. Even the most traditional mom-and-pop pizza makers in Italy partnered with food delivery platforms to cater to eaters in 2020.[2] Similarly, in Shanghai, grocery delivery apps became lifelines for citizens, farms, and food factory owners in spring 2022 under the extreme lockdowns.

Traditional companies may also benefit from collaborating with tech giants to access other resources, such as insights on customers. In central China, a tissue paper manufacturer, Corou, was in a business with a thin gross profit of about 3 percent. It initially faced several challenges, including lack of access to end consumers and supply chain issues. When it started to sell on PDD, China's second-largest e-commerce marketplace, Corou was able to have direct access to massive buyer traffic and received guidance on product features and pricing from the platform.[3]

Taking advantage of these resources, Corou implemented two key changes: first, the company slightly reduced the paper size, and second, it relocated its production line to the same industrial park as its raw material supplier, effectively reducing shipment costs. Just two years later, Corou achieved sales of an impressive 1.65 million orders through the PDD platform. It was also able to cut the price of the same product by 3.4 percent, as a result of production efficiency improvement and cost saving.[4]

Collaborating with tech giants also allows traditional businesses to access advanced technologies. For example, following OpenAI's decision to open its product to become a platform, companies across diverse industries began partnering with it and leveraging its expansive language model to empower their own businesses.

A vast majority of traditional businesses nowadays are engaging in collaborations with tech giants in one way or another. Many have seen some positive impact on their businesses. Smart rivals, however, don't just rest on these initial successes. Instead, they anticipate potential risks that could emerge down the line and begin designing defense strategies early in these partnerships. By proactively developing new capabilities to implement these defense strategies, these forward-thinking businesses strengthen their position and learn to effectively manage the complex frenemy relationships they hold with tech giants.

The Risks of Partnering with Tech Giants

As illustrated by the X Fire case, partners can turn into competitors. Tech giants can offer products and services that are competitors or substitutes for their partners' offerings. Such actions can erode the market share and profit margins of traditional businesses, diminishing their capacity to capture value. In essence, traditional businesses can find themselves in a paradoxical position where they are simultaneously partnering with and competing against the platforms.

Amazon Marketplace, launched in 2000, and open to third-party merchants, positioned the company as both a retailer and a platform provider. In the second quarter of 2022, third-party sellers accounted for a record 57 percent of all units sold on the platform.[5] Amazon has been reported to use the data it collects from these third-party merchants to develop its own private-label products, leveraging information such as total sales and the commission that the platform earned from each transaction.[6] Such competition should not come as a surprise. We are accustomed to seeing traditional retailers, from CVS Pharmacy to Costco Wholesale, put their own generic labels next to the brand-name products at lower prices on their store shelves.

Compared to such brick-and-mortar retailers, digital platforms enjoy a much lower cost of innovation and experimentation due to unlimited virtual shelf space. By inviting many third parties to list their products on their platforms, they are running a gigantic innovation lab where these third parties bear the cost of innovation and provide tech giants with insights on what products or services are likely to be successful.

In the case of offline retailers, running such contests among many products can be costly and difficult because of physical space constraints. In addition, the ability to aggregate a large amount of data within a short period gives tech giants more precise and early insights than offline businesses. Just as Steve Herbert Sr. told us, "I feel that, in many cases, I have done the market testing and development, have proven that a particular product will be successful and then Amazon comes in and 'steals' the business."

We have discussed a number of areas in which your company might need to be wary in dealing with tech giants. Here are some specific issues to consider as you establish partnerships:

Recommendation algorithms

Tech giants can design their recommendation algorithms to capture more value from their partners. The recommendation algorithms are often a black box to outsiders. Customers don't really know why some products are listed before others in their search results. In the United States, for example, in 2019 Amazon was reported to have changed its secret algorithm to prioritize profitability over relevance, or in other words, to favor its own brands that are more profitable to the company.[7]

Fees

Tech giants can raise fees once they have achieved market power to capture more value from third parties. During the Covid-19 pandemic, many restaurants in the United States had to pay hefty fees to food delivery companies, which they relied on to survive. While New York City enacted temporary caps on the commissions, three of the country's biggest food delivery giants, Door-Dash, Grubhub, and Uber Eats, filed a lawsuit in September 2021 to prevent the city from enforcing the fee-limit ordinance.[8]

Gatekeeper policies

As gatekeepers, tech giants can adjust policies that may not align with the best interests of third parties. For example, Apple announced a major policy (App Tracking Transparency) in 2020 that requires third-party apps to ask permission to track users' data. Positioned as a privacy-enhancing tool, Apple significantly reduced third-party apps' ability to provide personalized ads to their users without explicit user opt-in.

As a result, many app developers were faced with a decline in ad revenue that no longer allowed them to provide their apps for free and led them to change their ad-based business model to charge a fee.[9] Apple could benefit from this shift in business model as it frequently receives commissions from user payments.

Customer data

Working with tech giants is very likely at the cost of losing control of valuable customer data. Your company's access to customer data, and ultimately your relationships with these customers, is at the mercy of platform policies. For example, data access has been one of the major sticking points between restaurants and food delivery platforms. The food delivery platforms understandably want to own the customer relationship and retain that relationship. Restaurants feel that they provide the services and complain that they do not have access to eaters' information when orders are placed through the food delivery platforms. In New York City's attempt to help the struggling restaurants during the pandemic, it mandated the food delivery companies to share more data with restaurants, such as a customer's name, contact information, and delivery address.[10] DoorDash and Grubhub reacted by filing lawsuits against the city government in December 2021, citing customer data privacy and rights.[11]

In China, many consumers and merchants may still recall the rift between Cainiao Smart Logistics Network (Cainiao), the logistics arm of Alibaba Group, and SF Express, the country's leading courier, in 2017. Cainiao and SF Express had been close partners—SF Express shipped a significant portion of parcels for Cainiao. The dispute began when Cainiao announced that it would remove SF Express from its platform, which meant that Alibaba's e-commerce customers would not be able to choose SF Express as their courier. Cainiao cited data sharing issues as the reason for this decision. SF Express had stopped sharing certain data with Cainiao, which it believed was essential for smooth operations and providing better customer service. SF Express responded by accusing Cainiao of monopolistic practices and filed a complaint with the industrial regulator, China's State Post Bureau, claiming that Cainiao was using its dominant position in the market to force SF Express to share data.

The dispute caught sellers off-guard, scrambling to find other logistics options.[12] The standoff was later solved after the regulator stepped in and urged the two sides to find common ground and protect customer rights.[13]

Your value proposition

Another potential danger inherent in partnering with tech giants is related to the value proposition of your products.

Working with a tech giant can risk turning a product into a commodity, because these giants often provide a standardized environment that limits product differentiation and innovation. For example, e-commerce platforms use the same format to display each product. This uniform display may make premium products look similar to average products. In addition, platforms often have a recommendation system that favors low-cost and heavily discounted products, which can put pressure on businesses to lower their prices to remain competitive. This design choice can lead to a race to the bottom, where price becomes the primary differentiator and products become commoditized.

Let's Play Defense

Traditional businesses can use several strategies to mitigate the risks from such partnerships.

Decoupling

With so many potential risks ahead, some traditional businesses simply choose not to deal with digital platforms but play along by themselves. Nike, the world's largest sports company, in 2019 announced it would stop selling products on Amazon, a move to limit unauthorized sales as well as to redirect sales to its own online channels.[14] The US clothing retailer Gap, that sought multiple channels to boost its sales, ultimately decided not to sell on marketplaces such as Amazon, after a long internal debate.[15]

Similarly, as Amazon continued to enhance its in-house delivery capabilities and increasingly took on more shipping responsibilities, FedEx, a key enabler of Amazon's growth journey, decided to terminate its delivery agreements with the tech giant in 2019.

Such decisions are not easy to make. Clearly, companies, in their decision not to form partnerships, are knowingly forgoing the many advantages that partnerships bring. Even if they are willing to give up such benefits, they need to develop all the capabilities required to serve users directly on their own, and it is not always the case that they can do so at a lower cost than these tech giants could.

The tradeoff becomes even more challenging to assess when it involves new technologies and business models. Car manufacturers across the world face such a dilemma amid the race toward commercializing autonomous vehicles. From Waymo, Alphabet's self-driving unit, and Tesla in the United States to Huawei and Baidu in China, they all have expressed interest in partnering with traditional car manufacturers to develop autonomous vehicles. On the surface, there is a great deal of complementarity from this collaboration, and the partnership can be a win-win. Automakers excel in manufacturing cars but traditionally lack proficiency in AI. They do not possess digital assets such as high-precision maps that are critical for the driverless world. These tech giants have no expertise in car manufacturing.

But at the same time, forward-looking firms may have several concerns. In the driverless world, a substantial portion of value creation in car rides will stem from the services offered through the vehicle because passengers no longer have to pay attention to the road. For example, Google can deliver location-based advertising and ad-sponsored services like YouTube to passengers. These tech giants are poised to capture most of this value because of their control of the software layer. Should car manufacturers strive to secure a share of this value for themselves?

In addition, car manufacturers worry about the potential erosion of their brand value. With the development of the auto industry, the most critical and valuable aspects of automobiles in the future are their intelligent systems, content delivered on the cars, and autonomous capabilities. Tech giants will continue to retain these most cutting-edge and valuable aspects in their own hands. Car manufacturers might find themselves reduced to serving as mere laborers for these tech giants.

Moreover, when multiple car manufacturers operate on the same technologies provided by these tech giants, consumers may perceive less differentiation among the vehicles. In the future, a potential car buyer might ask whether they should opt for a Google-empowered or Tesla-empowered car as their primary consideration.

When asked about the possibility of partnering with Huawei and leveraging its autonomous driving technology in 2021, Chairman Hong Chen of China's SAIC Motor Corporation, one of the largest automobile manufacturers, responded: "Cooperating with third-party companies like Huawei on autonomous driving is not acceptable for SAIC. It's like having a company provide us with a complete technical solution, which would make them the soul and SAIC the body. SAIC cannot accept this outcome and wants to keep the soul in its own hands."[16]

Sharing similar concerns, major car manufacturers moved aggressively to develop in-house AI capabilities. For example, in 2016, General Motors, aiming to speed up self-driving car development, bought Cruise Automation, a San Francisco-based autonomous driving startup founded in 2013, for more than $1 billion.[17] SAIC, GM, Toyota, Mercedes-Benz, and Bosch all invested in Momenta, an autonomous driving solution provider from China. Toyota also engaged in research collaboration into AI with many top US universities, including MIT, Carnegie Mellon, Stanford, and Princeton. In March 2023 GM announced its decision to discontinue the use of Apple CarPlay and Android Auto on certain new electric vehicle models to have greater control of cars' dashboard display.[18]

We discussed Disney+ in chapter 1, an online streaming service that the company released in 2019 after ending its distribution deal with Netflix. The history and relationship between Disney and Netflix in terms of streaming can be characterized by both collaboration and eventual competition. The journey began in 2012 when the two companies reached a multi-year licensing agreement, allowing Netflix to stream Disney movies and TV shows.[19] However, as the streaming industry grew and Disney recognized the value of direct distribution, the company decided to disintermediate Netflix and launched its own streaming service. By leveraging its rich portfolio of content, which included popular franchises like Marvel, Star Wars, and Pixar, Disney left Netflix to compete with it directly.

Decoupling is an extreme way of managing the relationship with tech giants. With decoupling, traditional businesses position themselves as direct competitors instead of partners of tech giants. However, if your business is not strong enough to build all digital capabilities yourself like Nike and Disney, it is important not to forget about the friend element of frenemy relations when managing these relationships.

Strategic channel-bridging

Instead of cutting ties completely, a firm could consider a strategy to bridge channels. In this case, a firm simultaneously leverages the reach and resources of a tech giant while also working to incentivize users to directly engage with the firm, thus essentially bypassing the tech giant—disintermediation. Such a strategy reflects a complex and potentially delicate balancing act between utilizing available distribution channels and promoting direct consumer engagement.

Huazhu Group, China's second-largest hotel group by number of rooms, used this strategy to manage its relationship with online travel agencies (OTAs).[20] OTAs, one-stop platforms that enable travelers to search for and directly book hotels, flights, vacation packages, and other related services, have ruled China's hotel booking market for many years.

Like its peers, Huazhu has long relied on OTAs for new customer acquisition. In fact, its executive chairman and founder Qi Ji, cofounded China's biggest OTA, Trip.com Group, in 1999, six years earlier than he founded Huazhu, and he remains on the board of directors at the OTA today. Huazhu has been exploring multiple approaches to reduce its reliance on OTAs, an expensive distribution channel for hotel sales with commission rates of as much as 20 percent in China as of 2020.[21]

To implement strategic channel-bridging, Huazhu invested heavily in its own digital channels and endeavored to motivate guests to join its loyalty program and book directly through its own online and mobile channels. It had 193 million members at the end of 2021, making it the largest hotel loyalty program in China, with 11.9 percent as corporate members.[22]

When we visited a Huazhu hotel in Shanghai on a July afternoon in 2019, a receptionist was trying to convince a guest who had booked the room from Trip.com to become a Huazhu member. After purchasing an annual gold membership, the guest would enjoy a discount for the next booking via Huazhu's direct channel, with a free breakfast offered.

Huazhu president Xinxin Liu described the company's relationship with the OTAs as "coopetition." OTAs, with a massive user base, are still most travelers' first go-to places. "Thus, we are in a long-term battle of negotiation with the OTAs for better terms. We are promoting our direct sales and managing the OTA channels simultaneously," Liu told us.[23]

Similarly, while working with delivery platforms, many restaurants try to disintermediate food delivery platforms to connect with eaters directly. Many restaurants already have or can easily build delivery capabilities themselves; after all, delivery does not require skills beyond driving a car or riding a bike. White-label on-demand delivery services (such as Olo and Relay) have become popular. They allow restaurants to enable delivery through their virtual front doors (their websites and apps). Given this trend, it is not surprising that restaurants would slip notes or menus inside their food orders to encourage their customers to disintermediate food delivery platforms. In addition to fee avoidance, disintermediation allows participants to avoid sharing their transaction data. Firms can also strengthen their brands and have more control over some quality metrics by interacting with their customers directly and exercising better controls over transactions.

Preserving and enhancing your unique value creation

Traditional businesses should actively strategize to safeguard their uniqueness and develop innovative approaches to prevent tech giants from easily replicating their value propositions.[24]

Back in the United States, in the X Fire case, the owners made efforts to ensure that Amazon would not be able to source identical products from the same suppliers. They reached out to the suppliers directly and visited their CEOs in person. The Herberts presented a compelling argument to their suppliers, emphasizing the significance of their brick-and-mortar store, along with the strong customer relationships and trust they had cultivated. "People come in to ask us for product recommendations," said Steve Herbert Sr. "We told our suppliers if they were going to sell straight to Amazon, we would actively persuade our customers in our store not to buy their products, or we would stop selling those products in our store altogether."[25]

X Fire also educated the suppliers that as a top paintball seller on Amazon, it helps them get to Amazon's customers, even if they don't sell products to the tech giant directly.

Furthermore, X Fire made amendments to supplier contracts, introducing a return policy that granted them the right to return products if manufacturers opted to sell directly to Amazon. And X Fire indeed executed the contracts whenever violations occurred, hurting the suppliers. Consequently, X Fire effectively discouraged several of its largest suppliers from pursuing direct sales to Amazon.

Such strategies require persistent efforts. Industry dynamics and supplier consolidations can jeopardize the hard-earned agreements secured by X Fire. With each merger among suppliers, X Fire faces the task of persuading a fresh leadership at the helm once again.

Sometimes, collaborations with tech giants, when well designed, can help traditional businesses deepen their differentiation. Take Best Buy as an example. For years, Amazon's rapid growth and dominant position in e-commerce have posed a significant threat to consumer electronics stores. To stay relevant, in addition to price matching Amazon and growing its online presence, Best Buy doubled down on its key differentiation from Amazon: its in-store experience. Under the leadership of then-CEO Hubert Joly, the company launched training programs and new incentive programs for its store associates to empower them and make them more engaged.

Best Buy has expanded its service offerings, such as the Geek Squad, which provides technical support, installation at a customer's home, and repair services for consumer electronics. Best Buy has additionally secured agreements with major electronics firms such as Apple, Microsoft, Samsung, Apple, and Sony to create dedicated sections or mini-shops within their retail locations.

This store-in-store format allows customers to explore a wide range of products from these brands and receive specialized assistance from dedicated brand representatives.

But, in a surprising move, in 2018, Best Buy teamed up with longtime rival Amazon. Under the new partnership, Best Buy would sell smart TVs equipped with Amazon's Fire TV software and Alexa in its stores. Amazon faces a hurdle with TVs, as consumers frequently hesitate to buy without a firsthand look at the product. Assessing comparative quality solely from images on computer screens proves challenging. Certain features, such as high dynamic range (HDR), remain inaccessible for evaluation without a physical presence.[26] Amazon therefore needed a popular retail outlet for its TV business. By selectively partnering with Amazon, Best Buy further enhanced its brand image as a company offering superior in-store experiences, solidifying its competitive advantages over Amazon. The partnership also reduced Amazon's incentive to launch offline electronic stores by itself.

Domino's employed a similar strategy to manage its relationship with food delivery platforms. As a leading pizza company globally, Domino's attracted considerable interest from these digital platforms. As discussed in chapter 1, historically, Domino's opted not to collaborate with them. However, in a groundbreaking move, Domino's forged a global exclusive agreement with Uber in July 2023, enabling customers to order Domino's pizza through Uber Eats. Domino's asserted that this alliance would broaden its customer base, reaching new segments in thirteen international markets.[27] Crucially, Domino's distinguishes itself from many restaurants partnering with Uber by using its own drivers for delivering orders from Uber, thereby upholding its reputation as a delivery specialist.

We are entering an era where regulators from Brussels, to Washington, and Beijing are reining in the power of tech giants. While the motivation varies, the governments all agree that the small cadre of these large digital companies has become unprecedentedly expansive with an urgency to limit their influences.

Against this backdrop, with less power and limited strategic options, traditional businesses may turn to public opinion, lawmakers, or the courts to improve their relative position and bargaining power. For example, Apple's decision to lower its commission rate to small third-party businesses can be partially attributed to the legal dispute with Epic Games, an American video game and software development company with several popular game franchises such as Fortnite. The conflict began when Epic Games implemented its own in-app payment system to bypass Apple's 30 percent commission on in-app purchases in August 2020.

In response, Apple removed Fortnite from the app store, leading to a high-profile lawsuit filed by Epic Games, alleging that Apple was abusing its market power to charge a high commission and imposing restrictions against using alternative payment methods within iOS apps. In September 2020, Epic Games joined forces with thirteen other prominent companies— including the music streaming platform Spotify and Tinder owner Match Group—to establish the Coalition for App Fairness (CAF), which aimed to reach a fairer deal for the inclusion of their apps into the Apple App Store or the Google Play Store. As of 2021, CAF had grown to more than fifty members.[28]

Although the case was decided in Apple's favor, the lawsuit and surrounding controversy have sparked a broader conversation about the power and control that large tech companies like Apple have over app developers and digital marketplaces.

In November 2020, Apple introduced the App Store Small Business Program, which reduced the commission rate from 30 percent to 15 percent for developers earning less than $1 million in annual revenue from their apps. In addition, the court also rejected Apple's anti-steering policies, thus allowing iOS developers to provide information on other payment options to their users in the app so that users can complete payments elsewhere.

Epic Games possessed the financial resources to fight Apple and continues to perform well following the conflict. For many others in the industry, the outcome of this confrontation is crucial to their survival. Take news publishers as an example. More than two thousand local newspapers in the United States were shuttered between 2004 and 2020, eliminating half of all newspaper jobs.[29] (Some of our veteran journalist friends also lost their jobs during their publishers' recent cost-cutting layoffs.) The decline of this industry was partly because tech giants such as Facebook and Google attracted ad revenues away from them. At the same time, Facebook and Google both benefited from news content, which helped retain user engagement and generated more ad revenue, but they had low incentives to compensate news publishers.

Not long ago, tech giants such as Google had the upper hand in the battle with traditional media, despite the fact that publishers across western Europe pushed to introduce copyright legislation to force these big techs to pay licensing fees.[30] In

2004, Germany's biggest news publisher, Axel Springer, did not restrict Google from using its content, because the traffic on its sites plunged by 80 percent within two weeks after it stopped allowing the tech giant to publish its news snippets.[31] It had to back down. The tables turned in 2021 when the Australian parliament passed the landmark News Media and Digital Platforms Mandatory Bargaining Code, which aimed at protecting news publishers from the dominance of Google and a few other tech giants.[32] Despite Google's initial threat to pull its search engine out of the country, Australia's government stood firm, as the code requires tech platforms to negotiate and pay news publishers for their content. Google and Facebook ultimately struck deals with news organizations, with Google securing nineteen content agreements and Facebook securing eleven.[33]

These deals have been seen as a way to subsidize the struggling media industry and support journalism. Google's and Facebook's parents Alphabet and Meta have paid over AU$200 million in payments to news publishers during the code's first year of the operation.[34]

What the Australian government achieved has a ripple effect as other governments around the world follow suit.

Canada introduced similar legislation, while in France Google in 2022 agreed to a new slate of licensing deals with French publishers.[35] In the United States over two hundred American newspapers across several states filed antitrust lawsuits against Facebook and Google from 2020 to 2021, turning it into a national movement.[36]

US lawmakers unveiled a new version of the Journalism Competition and Preservation Act in August 2022.[37] If passed, it would allow small and medium newspapers to collectively negotiate with giant online platforms such as Facebook and Google at a level playing field.[38] Large publishers, such as *The New York Times* and the *Wall Street Journal*, won't qualify, preventing the bill from favoring big national publications.[39] As of December 2023, a vote was still pending.

As traditional businesses fight such battles, it is important for you to find strength in numbers. Although X Fire was a top paintball and airsoft seller on Amazon, it was still just one small David up against Goliath. One idea the company has thought of has been to find strength in numbers by working with other third-party sellers providing similar products and pressuring Amazon to stop sourcing products directly from suppliers when those products were already offered by third-party sellers on the site. With such a disparate group of sellers, however, X Fire quickly found that it didn't have the resources and ability to coordinate such a large effort.

In 2018, in response to concerns about the growing dominance of e-commerce platforms and their treatment of third-party sellers, small businesses in the United States formed groups such as the Online Merchants Guild to advocate for their interests. These groups, recognizing the need for a unified representation for online sellers, work collectively to address issues such as fees, intellectual property disputes, and competition with Amazon's own products. They aim to provide a unified voice for smaller sellers in negotiations with tech giants. Some of their efforts include fair and transparent platform policies, access to vital marketplace data, reasonable commission structures, and improved seller support and communication channels. These guilds also focus on fostering knowledge sharing and providing educational resources to empower smaller sellers with the information they need to navigate the complex digital marketplace landscape.

It's also crucial to understand that, despite the growing effectiveness of leveraging regulators or public opinion, as these examples demonstrate, the process to formulate and enact new regulations is complex and lengthy, and fraught with significant uncertainty due to the vast resources and influence wielded by these tech giants.

In addition, the regulators and traditional businesses may not always be able to make a successful legal case. Therefore, while traditional businesses may find it beneficial to consider this strategy as a defensive measure, they shouldn't depend solely on this approach for protection.

Things to Look Out For

The relationship between traditional businesses and tech giants is a dynamic one. As tech giants continue to gain industry-specific knowledge of traditional businesses and gain capabilities over time, their relative positions in the partnership may change and tension may rise.

As a result, even if traditional businesses have good relationships with tech giants today, they need to build capabilities to protect themselves as well as to compete against tech giants to stay relevant and thrive. When you foresee a future in which tensions might rise, it is important to start preparing for that future early.

For example, before FedEx cut ties with Amazon, it had already been positioning itself to provide delivery services for retailers like Target and Walmart and had reduced its business related to Amazon to just over 1 percent of its overall revenue. That made it far less painful to break up.

Car manufacturers may consider themselves fortunate, given that the development of self-driving vehicle technologies proved more challenging than anticipated. Even though Google (now operating under Waymo) began work on autonomous vehicles in 2009 with aspirations of public usage by 2017, we're still a considerable distance from a world populated by such vehicles. This delay affords car manufacturers valuable time to develop their capabilities to compete in this emerging landscape. Had autonomous vehicle technology become available much earlier, many car manufacturers would have been compelled to form partnerships with tech giants.

Developing effective strategies to manage frenemy relationships demands time. As a result, smart rivals anticipate potential tension and swiftly take action to arm themselves against the encroaching influence of tech giants, instead of passively observing how relationships unfold. An old adage is fitting here: the best time to plant a tree was twenty years ago, and the second-best time is *now*.

CHAPTER 6

Bounce Back from Disruption

Turn every setback into a comeback.

—Kobe Bryant

The fortune of New Oriental Education & Technology Group, once China's largest private tutoring service provider, changed overnight on July 24, 2021. On that day, the Chinese government's sweeping crackdown on for-profit after-school tutoring services sparked massive layoffs and share selloffs of tutoring companies. New Oriental also let go sixty thousand employees, with the company's market value shrinking 90 percent, some $28 billion, that year.[1] For its founder, Michael Yu, it was probably the biggest crisis that the company faced over the past thirty years.

Started as a provider of English tutoring in 1993, New Oriental had experienced rapid expansion from K–12 private tutoring and preschool education to book publishing and overseas study consulting.[2] It became the first Chinese educational institution to debut on the New York Stock Exchange in 2016, while its subsidiary Koolearn was the first online education service provider to hold an initial public offering (IPO) on the Hong Kong Stock Exchange.[3]

Faced with mounting pressure, Yu attempted to shift its business focus to sectors unaffected by regulations, such as dance classes, adult education related to vocational training, and even selling hardware educational products. But he had little success. Fortunately, in November 2021, Yu also steered the company into e-commerce by selling agricultural products online. The new business took off in June 2022 when its former tutors started teaching English during the live streaming shopping sessions. Teaching English for free and selling steaks and rice, an unlikely combo, went viral on Douyin, the Chinese version of TikTok. New Oriental, renowned for its skilled teachers, transformed its teaching expertise into marketing skills. The teachers-turned-influencers blend English teaching, humor, and personal anecdotes—reminiscent of their classroom style—into product sales.

Though New Oriental was hit by regulation rather than new technology, its experience is a familiar one: Its business was disrupted unexpectedly, putting its entire future in doubt. But when life gives you lemons, make (and sell) lemonade. Confronted with this existential disruption, Yu was able to create an altogether new business. Within two weeks after integrating English teaching into its live streaming e-commerce business, followers of the company's live streaming brand Oriental Select on Douyin rose nearly 20 times to 18.5 million and booked product sales of $76.7 million.[4] Oriental Select has been the top-ranked live streaming account on Douyin since June 2022. The company's share price soared almost 300 percent between July and December 2022, and it returned to profitability during the third quarter of the same year.

From classrooms to e-commerce, New Oriental found a new growth engine based on its core competency. It innovated in the live streaming e-commerce business, differentiating from both tutor providers and e-commerce platforms. The company's former tutors pose a stark contrast to other live streaming influencers, who are often hyping up products and urging people to buy fast and more.

Disrupted by government regulation, New Oriental bounced back and took advantage of the disruption: it picked a sector that has government policy support as it aligns with China's goal of shoring up the rural economy. Government policies, once a disruptor, became a propellant to the company.

Other private tutoring companies also tried to change course, with some pivoting into educational hardware or life-science businesses. These ventures, which seem to be far from their original core businesses, have been less successful than New Oriental.

Just like New Oriental, companies today live with the risk of disruption. In previous chapters, we outlined ways to reduce the likelihood of running into a crash course with tech giants. But disruption could still happen no matter how visionary your company is. When it happens, will you be able to bounce back like New Oriental?

In this chapter, we discuss approaches that traditional businesses can consider in order to turn around their businesses and rebound after disruption—and why standard approaches may fail.

Responding to Disruption

How should incumbent traditional businesses react to these disruptors? Unfortunately, the old school responses often don't work well. Incumbents typically either launch a fighting brand to go head-to-head, or try to buy the new entrant, or just try to focus on their core customers. But these strategies all have their flaws, as we'll see below. We will also explore two additional strategies.

Launching a fighting brand

Often, the natural reaction for many traditional businesses is to take on a disruptive tech entrant at its own game through launching a similar offering (fighting brand). When you are already not doing well, fighting is a possible but very challenging approach to turn your company around.

Fighting a disruptor requires willingness to suffer more pain. (This pain is independent of whether the traditional business tries to launch the fighting brand in-house or in a separate business.) Because of deteriorating financial performance, professional managers, without strong support within the organization, can be quickly ousted during this process. As a result, they are more likely to choose other options. Organizations become more willing to fight when they are still managed by the founders and are private, so they are not under pressure from Wall Street.

This also means that the best time to fight disruptors is when the core of traditional businesses is healthy and growing, and the additional profits can compensate the losses from fighting disruptors.

Indeed, by the time video rental giant Blockbuster launched its online business in 2004 to fight against Netflix's mail-order DVD service, which began in 1999, it was already too late. The company was already carrying around $1 billion in debt at the time.[5] Its shareholders and a few board members vehemently opposed the CEO's plans to expand Blockbuster's online business and eventually replaced him with someone who would refocus on brick-and-mortar stores to reduce loss. According to Netflix's CEO Reed Hastings, "If it hadn't been for their debt, they could have killed us."[6]

Fighting a digital disruptor with a similar offering takes more than the willingness to match their low prices and the willingness to cannibalize your existing business. As the story of Nigeria-based EbonyLife Media, discussed in chapter 3, illustrates, building a digital operation requires a new set of capabilities, many of which are not so easy for incumbents to acquire in a short period.

The traditional sector we belong to, higher education, is not immune to the disruption caused by technology, either. Our home institution, Harvard University, faces the rising tide of online learning platforms, such as Coursera, LinkedIn Learning, and Udemy. Coursera, founded just over ten years ago, now serves more than 100 million learners and offers over seven thousand courses from experienced practitioners and award-winning professors across numerous industries and universities, posing a significant challenge to traditional education models. So, how should Harvard respond?

Unlike Blockbuster, Harvard had the financial strength to grow a learning platform like Coursera. Harvard and MIT cofounded edX, an online learning platform, shortly after Coursera's launch in 2012. Yet, building advanced technological infrastructure, offering a wide course selection comparable to these platforms, and launching aggressive marketing campaigns to attract students are not within Harvard's specialty. The subsequent sale of edX, a loss-making business, to 2U, another ed tech firm, in 2021 serves as a cautionary tale.

Buying the disruptor

Sometimes companies may consider biting the bullet and acquiring the disruptor. It is often easier said than done. The PetSmart acquisition of Chewy is an example of a traditional business buying a disruptor, but it was not an easy deal.

The competition in the US pet product market has always been dog-eat-dog. Back in 2015, PetSmart owned fourteen hundred brick-and-mortar pet stores in the United States.[7] When PetSmart noticed Chewy, an online retailer of pet-related products, eating up its market share, it decided to purchase the startup instead of competing with it head-to-head. Shoring up PetSmart's small e-commerce site to beat Chewy would have been too costly in both money and time. Uncertainty also persisted regarding PetSmart's ability to lure the essential Silicon Valley–caliber talent required for the success of its e-commerce business.

In 2017, PetSmart acquired Chewy for $3.35 billion, the largest e-commerce deal in history at the time. The two companies expected the deal to enhance both entities' capabilities and reach.[8] However, the deal was soured by PetSmart's lenders and the broader market: Chewy, just like most disruptors in their early days, was not profitable, and PetSmart loaded $2 billion in debt to its already high $6 billion debt for an unprofitable rival.[9]

After the acquisition, the sales of PetSmart slowed, while the Chewy losses only widened. PetSmart's bonds once fell to a level that signaled the company was heading for bankruptcy. Its CEO stepped down four months after the acquisition.

BC Partners, the private equity firm that had bought PetSmart in 2015, stepped in to help restructure Chewy's ownership, paving the way for an IPO of Chewy. It also revamped PetSmart's strategy to emphasize its core capabilities of offline pet services such as veterinary clinics and grooming, which couldn't be matched by its online counterparts.

Chewy went public in June 2019, and its share price jumped nearly 20 percent on the first day of trading, resulting in a paper gain of about $10 billion for PetSmart.[10] As more pet owners shifted to shopping online during the Covid-19 pandemic, Chewy's business took off on the path to profitability. It led online stores in pet supplies in the United States by a large margin over PetSmart in 2022, while PetSmart, with 1,669 locations, remained the number one pet specialty chain in North America.[11]

While PetSmart's acquisition of Chewy was a great financial investment, the integration of the two companies did not go smoothly. Some insiders noted that Chewy provided little assistance to PetSmart regarding its online strategy.[12] At the same time, Chewy's growth cannibalized PetSmart's brick-and-mortar sales. PetSmart announced plans to sell all shares of Chewy around October 2020. BC Partners also decided that the two companies would be better off on their own and planned to split them apart.[13]

PetSmart's experience highlights a fundamental challenge with acquisition strategies: acquiring a disruptor does not resolve the underlying issue of business cannibalization between the two entities.

Retaining your core customers

Traditional businesses sometimes choose to double down on their existing customers and try to retain them in the face of disruption.

Marriott versus Airbnb represents a familiar competitive dynamic between digital entrants and traditional businesses taking place in many markets today. Marriott is the world's largest hotel company with thirty brands (ranging from midscale to luxury). Across its brands, it has more than 1.5 million rooms in 138 countries and territories.[14] It also operates a popular Bonvoy loyalty program, with over 200 million members at the end of 2023.[15] Marriott has recovered successfully from the Covid-19 pandemic and has benefited from the surge in travel after the pandemic.

Airbnb, an online marketplace that enables hosts to rent out their rooms to travelers, posed a threat to Marriott. Airbnb does not own or manage any accommodations. It started in 2008 after its cofounders began renting mattresses in their apartment to make a few bucks. Many hosts started to share their unused property, often an extra bedroom or couch, to earn some income. Over the years, Airbnb has grown exponentially, attracting a variety of properties. It offered more than seven million listings on its marketplace at the end of 2022, with fourteen thousand new hosts joining the platform each month that year.[16] Airbnb's market cap was 40 percent more than Marriott's as of April 2024.

Unlike Marriott, which conducts inspections to ensure that its properties meet its quality standards, Airbnb relies primarily on consumer reviews and algorithms to control the quality of listings and personalize its offerings to guests. For example, one of its recent efforts involves using AI to prevent house parties.[17] Still, guests are more likely to have quality and safety concerns with Airbnb stays than Marriott hotels.

Initially, Airbnb primarily focused on leisure stays rather than business travel. In contrast, business travel constituted 80 percent of Marriott's room nights. As a result, the impact of Airbnb typically concentrates on Marriott's low-end brands, whereas travelers substitute hotel stays with Airbnb for leisure travel. But Airbnb has identified business travel as a major growth area. Airbnb hosts today can find many tips online on how they can better attract business travelers.[18]

Cognizant of the different capabilities each firm has, instead of offering a fighting brand, Marriott sought to retain its core customers (business customers or customers who care about standard and reliable quality of service) through several new offerings. For example, it created a private-home rental service called Homes & Villas in 2019, which had grown into a listing of sixty thousand properties across seventy-five countries by

2022. Unlike Airbnb, Homes & Villas partners with select professional home management companies to ensure that properties meet the highest standards.

In November 2022, Marriott responded to the rising consumer demand for accommodations that combine work and leisure by introducing Apartments by Marriott Bonvoy. This offering is also in the upper-upscale and luxury segment properties that are managed by Marriott.

These offerings make Marriott's high-end customers and business travelers happier and thus help slow down Airbnb's encroachment. For example, 90 percent of bookings for Homes & Villas come from Bonvoy members. Travelers accumulate and redeem Marriott points in the Bonvoy loyalty program when they stay at Homes & Villas or Apartments. These offerings don't, however, directly attract Airbnb's primary customers—those who are satisfied by a good-enough stay and do not want to pay high prices for high quality. As a result, this strategy, which focuses on retaining core customers, buys Marriott time but does not eliminate the threat.

A more recent offering by Marriott, StudioRes, designed for the affordable midscale extended stay tier, has started to allow it to tap into the more budget-conscious market segment Airbnb serves.[19]

Traditional businesses should leverage the extra time afforded by such new offerings to prepare for more significant innovations. Ultimately, a company like Marriott must identify strategies to reduce costs (and thereby prices) and enhance personalization, all while upholding its consistently high-quality standards. This approach will enable it to reclaim many budget-conscious customers from competitors like Airbnb. Focusing on core customers is seldom a sufficient response on its own.

Lobbying for a level playing field

Airbnb serves as a platform that connects travelers with hosts who rent out rooms to travelers. But most people aren't aware that Marriott plays a similar role. Marriott only owns a small fraction of its properties. The majority of Marriott's hotels are operated under franchise or management agreements with third-party property owners. While Marriott is not primarily a digital platform, a significant portion of its business model is based on a platform-like approach, attracting property owners to adopt the Marriott brand and connecting them with its users.

This model is very similar to Airbnb, as both companies need to attract and cater to two distinct groups of customers. As our colleague Chiara Farronato's research shows, the key differences between Marriott's and Airbnb's models are how they manage quantity and quality. With home-sharing platforms like Airbnb, a host can list a room or an apartment for rent in several minutes.

Consequently, when demand fluctuates due to factors such as seasonality or significant events in a city, more hosts may list their properties, motivated by the higher earning potential. In contrast, the capacity for Marriott is relatively fixed, and expanding it requires property owners to make nontrivial investment—and such investment is not easily reversed.

A second distinction between the operating models of traditional firms and digital-native companies lies in how they ensure quality. Marriott operates with a set of quality tiers, ranging from economy to luxury. Within each tier, every hotel needs to adhere to a set of quality standards. Marriott ensures its locations uphold these standards through regular inspections, whereas Airbnb mainly depends on guest feedback to monitor the quality of its listings.

These key differences capture different operational choices, with Marriott exerting more control over the properties it works with. Airbnb also aspires to have full control of travelers' experience during their stays. For example, the mission of Airbnb, according to its 2019 Business Update, is to "create a world where anyone can belong anywhere and we are focused on creating *an end-to-end travel platform that will handle every part of your trip*" (emphasis added).[20]

Given the similar objectives between the digital-native company and the traditional business, it is not surprising that hotels increasingly demand that Airbnb properties be regulated in the same way as hotel properties. The American Hotel and Lodging Association (AHLA), a trade association representing the US hotel and lodging industry, has lobbied for stricter regulations on these platforms to ensure a level playing field and address various concerns associated with short-term rentals.

Cities around the world from San Francisco, to Paris, London, Amsterdam, and Barcelona started to impose requirements on short-term rental properties, just like on hotels. For example, in San Francisco, hosts of short-term rentals must go through the process of registering with the city and obtaining a business license. Additionally, they are obligated to comply with local safety regulations, which include the maintenance of functional smoke detectors, carbon monoxide detectors, and fire extinguishers.

Similar to hotels, these hosts are also subjected to a 14 percent transient occupancy tax. Like San Francisco, Amsterdam enforces regulations that require hosts to register with the city, while also limiting the number of days for short-term rentals of entire homes to thirty days per year. Hosts in Amsterdam are also subject to a tourist tax, mirroring the payments made by hotels. Furthermore, the city has implemented safety and health regulations to ensure the well-being of short-term rental guests. These measures aim to alleviate the cost disadvantages faced by hotels and foster a more equitable environment for the traditional hotel industry.

Disruptors often position themselves as technology companies to avoid regulation and achieve high valuations. However, in practice, many aim to create or aspire to similar value propositions as traditional businesses.[21] Traditional businesses need to be diligent to educate regulators on the identity of these tech giants when they are practicing the same activities, thus leveling the playing field to ensure fair competition.

As another example, Ant Group started with the creation of Alipay, China's leading mobile payment platform, in 2004 for its parent company Alibaba's e-commerce platforms, primarily Taobao and Tmall. Driven by the trust-building mechanism of escrow payments, where funds were held until buyers confirmed satisfactory receipt of goods, Alipay quickly gained popularity in China, revolutionizing the way people made transactions online.

Over time, Alipay expanded its services beyond online payments. In 2013, it introduced a money market fund called Yu'e Bao, allowing users to invest their spare cash and earn interest. This offering proved immensely popular, attracting millions of investors and growing into one of the world's largest money market funds.

In 2014, Alipay transformed into Ant Financial Services Group by focusing on leveraging technology to provide inclusive financial services to underserved individuals and small businesses. At that time, Alipay had 190 million active users and dominated mobile payments in China, capturing 82 percent of the market.[22] It introduced Sesame Credit, a credit-scoring system based on user data, which provided a convenient way for individuals to access credit services. Ant also held 30 percent of the shares of MYbank, discussed in chapter 2, focusing on loans for SMEs. Additionally, it expanded into wealth management, insurance, and lending, offering a comprehensive suite of financial products and services. Its lending platforms Huabei and Jiebei allowed users to borrow money for online purchases or other expenses without needing a credit card.

The company's mission is to "enable all consumers and small businesses to have equal access to financial and other services through technology."[23] At first glance, this mission statement does not differ much from that of banks, as banks use technology, too. To set itself apart from banks, Ant Financial in 2020 renamed itself Ant Group and referred to itself as a techfin instead of a fintech company to emphasize its tech identity and to avoid being considered as a financial institution.

This strategy did not work out in the end. While Ant's approach "is not to take credit risk" itself, and only funds 2 percent of each loan, it has exposed the risks of China's financial system, because most of the company's 1.73 trillion yuan ($263 billion) debt came from Chinese banks and trusts as of the end of June 2020.[24] Chinese regulators were especially concerned about the weak risk management of those smaller and rural banks that supplied loans to Ant's customers.[25]

Ant Group's highly anticipated IPO was called off by the Chinese government just two days before its listing on the Hong Kong and Shanghai stock exchanges in November 2020. And a few days earlier, in October, billionaire founder Jack Ma criticized China's financial regulatory system in a speech at the Bund Summit in Shanghai, suggesting that it stifled innovation and hindered the growth of fintech companies. Ma's remarks drew widespread attention and raised concerns among Chinese regulators as challenging the authority and credibility of the regulatory system. Besides halting its IPO, Chinese regulators also intensified scrutiny of Ant Group's operations and compliance with financial regulations, which ended with a 7.12 billion yuan ($984 million) fine for the company in 2023.[26]

As digital disruptors emerge in many traditional industries, it is important for traditional businesses to pay attention to the activities these disruptors are actually performing. When they engage in similar activities as you do, you need to educate regulators as to why they need to be regulated the same to achieve a level playing field. Regulation and legislation generally lag the constantly evolving technology and it takes time to educate regulators on the identity of new disruptors. It is thus important for traditional businesses to work together and be persistent in educating the regulators.

Crafting a new path

There is one additional response to disruption that is consistent with the overall advice of this book. Rather than going head-to-head, traditional businesses should consider their core capabilities and chart a new path based on what they do best.

Unlike Harvard University, which chose to build a digital learning platform similar to Coursera, Harvard Business School (HBS) chose a different path for growth. Known for its unparalleled expertise in participant-centered, case-based learning, HBS opted to leverage digital technology to expand its unique learning model from traditional classrooms to an online environment, rather than merely mimicking Coursera. More crucially, HBS aimed to deliver the best possible participant-centered learning experience online by incorporating online sessions into existing programs and introducing numerous new online courses. Digital technologies facilitated increased interaction among its students, and even those hesitant to participate in physical classrooms became more engaged in its online learning environment. Reflecting on this journey, our colleague Stefan Thomke, chair of the school's General Management Program, shared his view with us in 2023:

> Potential program applicants initially assumed that our
> live online classes were inferior to in-person learning. They

couldn't imagine what we were offering and didn't know the many experiments we had run to develop a transformational learning experience online. In the latest iteration, we combined the best of online and in-person learning. Once participants experience our program, we receive no pushback on program fees, which have remained the same.

This new strategy enabled HBS to strengthen and amplify its existing competitive edge in participant-centered learning, making it inimitable for other ed tech platforms. While HBS may not offer the extensive range of courses found on these platforms or attract a comparable volume of students, its online programs quickly became profitable after their launch.

Let's take a look at how Kodak's long-term competitor embraced disruption. The world's film business had been animated by the duopoly of Kodak and Fujifilm for decades. Facing the same disruption that Kodak did, Fujifilm had a radically divergent fate. Fujifilm focused on its core capabilities rather than on the markets it was in.[27] The company's R&D team took a year and a half to figure out which of their existing in-house technologies could match the future market. Fujifilm adapted technologies and capitalized on them in areas such as pharmaceuticals, cosmetics, and highly functional materials. In 2017, less than 1 percent of Fujifilm's profits came from traditional film photography products.[28] If Kodak had also set a new course and innovated, it might have survived and thrived.

Then there's the story of Garmin, the once market leader of GPS products for military and automotive industries. The company faced disruption from the GPS-enabled iPhone and Android phones, which came with free map apps. Mobile phone data was expensive in the early 2000s. However, with technology development, the cost of mobile phone data declined, and the need for a separate navigation device became obsolete. Within three years, sales of Garmin's GPS units plummeted by almost $1 billion, with 90 percent of the company's market value evaporated.[29]

In early 2009, Garmin announced plans to manufacture a location-specific cellular telephone in cooperation with Asus under a new brand, Garmin Nüvifone. After releasing a few phone models for the Windows Phone and Android operating systems, in 2010, Garmin stopped making smartphones. Instead, Garmin looked inward and mapped out a new direction based on its core capabilities in GPS technology. It invested in and expanded the technology into sports wearables, activity trackers, navigation for boats and airplanes and other segments. Over the years, the company differentiated itself through innovation, as its products are known for being waterproof and having long battery life. Their devices can track a wider range of sports than the competitors, from surfing to skiing. Garmin's share prices surged nearly eleven times from the low in February 2009 to an all-time high in August 2021.

Even Nokia eventually realized the wisdom of this strategy. Losing the smartphone race devastated Nokia but didn't kill it completely. The sale of its mobile unit to Microsoft in a way relieved the company from having to struggle with a declining business and allowed it to craft a new growth path.

Nokia bought back the joint venture Nokia Siemens Networks (NSN) funded by the proceeds of the Microsoft-Nokia deal.[30] After a few tough years, the company changed its operation strategy to refocus on its mobile networks and infra-structure business, an area where Nokia has significant exper-tise. The company rode the wave of 5G technology and held a 20 percent market share of the telecom equipment market worldwide (excluding China).[31]

Things to Look Out For

Each of these responses takes companies into very different directions, and as a result, highlights the need for careful stra-tegic thinking. There is no one right way to respond to dis-ruption, and the decision will obviously depend on traditional businesses' capabilities as well as growth opportunities in other markets.

Traditional businesses should also consider how rapidly the disruption is taking place. Occasionally, disruptions take a long time to cause significant harm to incumbents. When that's the case (such as the development of autonomous vehi-cles discussed in chapter 5), fighting becomes a more feasible strategy.

However, many digital disruptions take place quite quickly (such as ridesharing services of Uber and app-based iOS ecosystem) because network effects fuel their growth process, because disruptors are often willing to take a loss to quickly establish their positions, or because the disruptors are giants in another industry and have the resources to drive rapid adoption.

In general, when disruption is rapidly emerging and is hard to fight, traditional businesses are likely to find more success by actively exploring other growth opportunities.

The experiences of Nokia, Fujifilm, Garmin, and New Oriental are not unique. Even tech giants have used this approach to rebound. Apple, instead of struggling to grow its market share in the personal computer market against the more open Wintel architecture—the collaboration between Microsoft Windows and Intel to produce personal computers—decided to move into consumer electronics and smartphones. Microsoft, concerned that Apple's iPad and smartphone might disrupt its Windows business, launched Surface and Windows Phone to compete, but neither managed to turn the company around.

Microsoft finally found a new growth opportunity with Azure, its cloud service. The cloud business took off successfully, partly because it built on Microsoft's strength in software development and server technologies, as well as its relationships with many corporations as a result of its Windows and Office businesses. Microsoft integrated Azure with Windows and Office—a strategy that not only generated new growth opportunities, but also helped enhance the defensibility of the Windows and Office businesses.

While disruptions are difficult to predict or preempt, smart rivals demonstrate the ability to recognize such disruptions as they emerge. They strategically engage in battles only when the odds of triumph are in their favor. In situations where disruption unfolds swiftly and becomes inevitable, smart rivals rebound by forging a fresh growth trajectory that capitalizes on their existing proficiencies.

Become Smart Rivals

D igital technologies, including recent advancements in AI, provide vast opportunities for businesses to scale up and broaden their scope using data, algorithms, and networks. This is true for both tech giants and traditional businesses.

Many traditional businesses have found success by adopting a digital-first mindset, placing digital innovation at the core of their strategies. This shift is evident in how their leaders define their companies. For instance, Patrick Doyle, former CEO of Domino's Pizza, referred to his company as a tech firm that happens to sell pizza.[1] John Deere, founded in 1837 and known for its agricultural, construction, and forestry equipment, considers itself a tech company that produces awesome tractors.[2] Similarly, after observing that his company had twice as many engineers as bankers, Piyush Gupta,

CEO of DBS Bank, described his organization as a technology company that offers financial services.[3]

However, adopting a digital-first mindset alone is not enough. The experiences of these companies also underscore the critical need for selecting the right strategic direction, the central theme of this book.

Throughout this book, we have emphasized that traditional businesses should strive to become smart rivals by forging their own paths, steering clear of the pitfall of mimicking tech giants—often the most challenging route to growth and success. Becoming smart rivals entails designing unique strategies to amplify your competitive advantage, drive customer centricity, grow platforms and ecosystems, navigate frenemy relations, and adapt to disruptions. Business leaders must cultivate a smart rival mindset, consistently engaging in strategies that tech giants find difficult to compete against.

Now that you've read this book, it's time to focus on your own organization and consider how you can become a smart rival in your field. Begin by applying the insights from the book to answer the following questions:

- What are your company's *competitive strengths* in the digital age? How can you leverage digital technologies or other innovations to enhance these strengths?

- What *unique data* can your company collect, and how can its insights be distinguished from those compiled by tech giants?

- Considering your user base and data, is there an opportunity to develop a platform? What changes in your operating model are necessary to grow this *platform business*?

- How can you take advantage of tech giants to expand and strengthen *your ecosystem*?

- What potential risks do you foresee in *working with tech giants*, and what are your strategies for defense?

- What *disruptions* are taking place in your industry, and what strategies do you plan to implement in response?

As you ponder these questions, you'll see that the road to becoming a smart rival is paved with innovation. It involves using digital technologies to uncover solutions and opportunities that play to your unique strengths and set you apart from tech giants. This book aims to inspire and empower you to start your journey with enthusiasm and a clearer vision for the future.

The digital age is brimming with opportunities, waiting for smart rivals like you to seize them. Go get them!

Chapter 1

1. James F. Peltz, "Domino's Pizza Stock Is Up 5,000% since 2008. Here's Why." *Los Angeles Times*, May 15, 2017, www.latimes.com/business/la-fi-agenda-dominos-20170515-story.html.

2. Michael Janofsky, "Domino's Ends Fast-Pizza Pledge after Big Award to Crash Victim," *New York Times*, December 22, 1993, https://www.nytimes.com/1993/12/22/business/domino-s-ends-fast-pizza-pledge-after-big-award-to-crash-victim.html.

3. "Domino's Launches Revolutionary Customer Tool: Pizza Tracker," Domino's press release, January 30, 2008, https://dominos.gcs-web.com/news-releases/news-release-details/dominos-launches-revolutionary-customer-tool-pizza-trackertm.

4. "Domino's Launches Revolutionary Customer Tool."

5. Domino's Pizza, "Innovations," https://biz.dominos.com/about-us/innovations/.

6. Susan Berfield, "Domino's Atoned for Its Crimes against Pizza and Built a $9 Billion Empire," *Bloomberg Business Week*, March 15, 2017, https://www.bloomberg.com/features/2017-dominos-pizza-empire/.

7. Julianne Pepitone, "Domino's Tests Drone Pizza Delivery," CNN Business, June 4, 2013, https://money.cnn.com/2013/06/04/technology/innovation/dominos-pizza-drone/index.html.

8. Lizzy Alfs, "What's Next for Domino's Pizza? CEO Patrick Doyle Outlines Some Goals," *Ann Arbor News*, September 1, 2013, http://www.annarbor.com/business-review/whats-next-for-dominos-pizza-ceo-patrick-doyle-outlines-some-goals/.

9. "Domino's and Nuro Launch Autonomous Pizza Delivery with On-Road Robot," Domino's press release, April 12, 2021, https://dominos.gcs-web.com/news-releases/news-release-details/dominosr-and-nuro-launch-autonomous-pizza-delivery-road-robot.

10. "Domino's Pizza Inc (DPZ) Q4 2020 Earnings Call Transcript," The Motley Fool, February 25, 2021, https://www.fool.com/earnings/call-transcripts/2021/02/25/dominos-pizza-inc-dpz-q4-2020-earnings-call-transc/.

11. "Domino's Introduces a New Way to Order Using Uber Eats Marketplace," Domino's press release, July 12, 2023, https://

ir.dominos.com/news-releases/news-release-details/dominosr-introduces
-new-way-order-using-uber-eats-marketplace.

12. Matthieu Quenard, "Is Amazon Winning as a Prestige Beauty
Retailer?" LinkedIn, February 17, 2017, https://www.linkedin.com
/pulse/amazon-winning-prestige-beauty-retailer-matthieu-quenard/.

13. Quenard, "Is Amazon Winning?"

14. Rina Yashayeva, "Battle of the Beauties: Amazon vs. Sephora,"
Stella Rising, May 20, 2020, https://www.stellarising.com/blog/amazon
-premium-beauty-vs-sephora.

15. Sephora, "Virtual Artist," https://www.sephora.my/pages/virtual
-artist.

16. "Sailthru's Fourth Annual Retail Personalization Index Highlights
Thriving Retail Brands," Sailthru, March 3, 2021, https://www
.globenewswire.com/news-release/2021/03/03/2186321/0/en/Sailthru-s
-Fourth-Annual-Retail-Personalization-Index-Highlights-Thriving-Retail
-Brands.html.

17. Julianna Wu, "Carrefour China Owner Suning Reports 17% Drop in
2019 Net Profits," KrASIA, March 17, 2020, https://kr-asia.com/key-stat
-carrefour-china-owner-suning-reports-17-drops-in-2019-net-profits.

18. "Suning Finance Overview," company profile, PitchBook, https://
pitchbook.com/profiles/company/223658-29; Ant Group, "Our History,"
Ant Group, https://www.antgroup.com/en/about/history.

19. "Alibaba Buys Half of Guangzhou Evergrande Football Club," BBC
News, June 5, 2014, https://www.bbc.com/news/business-27709641; Shen
Xinyue and Denise Jia, "In Depth: How Suning Fell into Crisis as JD.com
Surged," Caixin Global, March 24, 2021, https://www.caixinglobal
.com/2021-03-24/in-depth-how-suning-fell-into-crisis-as-jdcom
-surged-101679296.html; Adam Jourdan, "China's Suning Buying Majority
Stake in Inter Milan for $307 Million," Reuters, June 5, 2016, https://www
.reuters.com/article/us-soccer-inter-milan-suning-idUSKCN0YR03T.

20. Shen Xinyue et al., "What Will Suning Have to Sell Next?" Caixin
Global, March 22, 2021, https://www.caixinglobal.com/2021-03-22/cover
-story-what-will-suning-have-to-sell-next-101678279.html.

21. Xinyue et al., "What Will Suning Have to Sell Next?"

22. "Alibaba, JD.com Lead in China, but a Few Others Are Making
Dents, Too," eMarketer, July 2, 2019, https://www.emarketer.com
/content/alibaba-jd-com-lead-in-china-but-a-few-others-are-making
-dents-too.

23. Blake Schmidt, "Billionaire Chairman Quits after Suning's China
-Led Bailout," Bloomberg, July 12, 2021, https://www.bloomberg.com
/news/articles/2021-07-12/suning-s-billionaire-chairman-quits-after
-china-led-bailout.

24. Katie Robertson, "The New York Times Passes 10 Million
Subscribers," November 8, 2023, https://www.nytimes.com/2023/11/08
/business/media/new-york-times-q3-earnings.html.

25. "The New York Times Company Reports 2021 Fourth-Quarter and Full-Year Results," press release, February 2, 2022, https://s23.q4cdn.com/152113917/files/doc_news/2022/02/NYT-Press-Release-12.26.2021-PpCb082.pdf.

26. "The New York Times Company Reports 2021 Results."

27. Katie Robertson and John Koblin, "The New York Times to Disband Its Sports Department," *New York Times*, July 10, 2023, https://www.nytimes.com/2023/07/10/business/media/the-new-york-times-sports-department.html.

28. "The New York Times Company Reports Fourth-Quarter and Full-Year 2022 Results," press release, February 8, 2023, https://s23.q4cdn.com/152113917/files/doc_news/2023/NYT-Press-Release-Q4-2022-Final-gRYq6MI.pdf, 13.

29. "Our Heritage," IKEA, https://www.ikea.com/us/en/this-is-ikea/about-us/our-heritage-pubde78e100.

30. Thomas Stackpole, "Inside IKEA's Digital Transformation," hbr.org, June 4, 2021, https://hbr.org/2021/06/inside-ikeas-digital-transformation.

31. Sarah Whitten, "Disney Expects to Take a $150 Million Hit as It Cuts Ties with Netflix—and That's OK," CNBC, February 5, 2019, https://www.cnbc.com/2019/02/05/disney-expects-to-take-a-150-million-hit-as-it-cuts-ties-with-netflix.html.

32. Jennifer Maas, "Disney+ Tops 150 Million Subscribers, Streaming Loss Narrows to $387 Million," Variety, November 8, 2023, https://variety.com/2023/tv/news/disney-plus-subscribers-150-million-earnings-1235784850/.

Chapter 2

1. Xiaoxiao Ma, "Belle Launched 3D Foot Scanner with Epoque [in Chinese]," Sina, August 15, 2018, https://news.sina.com.cn/c/2017-08-16/doc-ifyixcaw4988908.shtml.

2. John Kang, "Founding Family of Footwear Giant Belle Takes a Big Step into Hong Kong's Startup Scene," *Forbes*, December 22, 2021, https://www.forbes.com/sites/johnkang/2021/12/22/founding-family-of-footwear-giant-belle-takes-a-big-step-into-hong-kongs-startup-scene/.

3. Qingteng University, "Belle International [in Chinese]," Qingteng One Question, November 30, 2020, https://mp.weixin.qq.com/s/ziHdvk25Ci20CvDBa3XBxg.

4. Peter F. Drucker, *The Practice of Management* (New York: Harper & Row, 1954), 37.

5. Blake Morgan, "A Global View of 'The Customer Is Always Right.'" *Forbes*, September 24, 2018, https://www.forbes.com/sites/blakemorgan/2018/09/24/a-global-view-of-the-customer-is-always-right/.

6. Marty Swant, "The World's Most Valuable Brands," *Forbes*, August 30, 2023, https://www.forbes.com/powerful-brands/list/.

7. "Coca-Cola Freestyle," Coca-Cola Company, https://www.coca-colafreestyle.com/.

8. Megan Leonhardt, "Coca-Cola Is Offering $10,000 If You Can Create the Best New Drink Flavor," CNBC, June 3, 2019, https://www.cnbc.com/2019/05/31/coca-cola-is-offering-10000-if-you-create-the-best-new-drink.html.

9. Brian Trelstad, Nien-hê Hsieh, Michael Norris, and Susan Pinckney, "Patagonia: 'Earth Is Now Our Only Shareholder,'" Case 323-057 (Boston: Harvard Business School, March 2023).

10. Cara Salpini, "Patagonia Opens First Worn Wear Store," *Retail Dive*, November 18, 2019, https://www.retaildive.com/news/patagonia-opens-first-worn-wear-store/567533/.

11. Chelsea Batter, "Patagonia's Worn Wear Collection Is Saving the Planet," The Manual, April 21, 2020, https://www.themanual.com/outdoors/oatagonia-worn-wear-collection-recycled-recommerce/.

12. Gretchen Salois, "Patagonia's Cyber Monday Resale Success: 70% of Customers New to Worn Wear," *Digital Commerce 360,* December 15, 2022, https://www.digitalcommerce360.com/2022/12/15/patagonias-cyber-monday-resale-success-70-of-customers-new-to-worn-wear.

13. Salpini, "Patagonia Opens First Worn Wear Store."

14. "Quarterly Retail E-Commerce Sales," US Census Bureau News, August 17, 2023, https://www.census.gov/retail/mrts/www/data/pdf/ec_current.pdf.

15. "MYbank Aims to Bring Inclusive Financial Services to 2,000 Rural Counties by 2025," *BusinessWire*, April 30, 2021, https://www.businesswire.com/news/home/20210430005190/en/.

16. "How to Apply for a Loan [in Chinese]," MYbank, https://mobilehelp.mybank.cn/bkebank/knowledgeDetail.htm?id=1445#toMainCa.

17. "Alibaba Leverages Customer-Centric Strategy to Stay Ahead in China's E-commerce Race," KrASIA, June 20, 2022, https://kr-asia.com/alibaba-leverages-customer-centric-strategy-to-stay-ahead-in-chinas-e-commerce-race.

18. International Finance Corporation, "MYbank's Gender-Driven Approach to Lending," MYbank, August 2020, https://documents.worldbank.org/en/publication/documents-reports/documentdetail/150781614010889824/mybank-s-gender-driven-approach-to-lending.

19. Engen Tham and Paul Carsten, "Alibaba Affiliate Launches Internet Bank for SMEs, 'Little Guys,'" June 25, 2015, https://www.reuters.com/article/ctech-us-alibaba-banking-idCAKBN0P50WF20150625; Yu Sho Cho, "How AI and Vast Data Support Ant Group's Financial Empire," November 2, 2020, https://asia.nikkei.com/Business/Finance/How-AI-and-vast-data-support-Ant-Group-s-financial-empire.

20. "MYbank Aims to Bring Inclusive Financial Services."

21. Portions of this section are adapted from Feng Zhu, Anthony K. Woo, and Nancy Hua Dai, "Ping An: Pioneering the New Model of 'Technology-Driven Finance,'" Case 620-068 (Boston: Harvard Business School, April 2020, revised November 2020).

22. OneConnect Financial Technology Co. Ltd., "Micro-Expression Remote Interview System," OneConnect, https://www.ocft.com.sg/wp -content/uploads/2021/01/Micro-Expression-Remote-Interview-System -03.pdf.

23. Feng Zhu, Anthony K. Woo, and Nancy Hua Dai, "Ping An: Pioneering the New Model of 'Technology-Driven Finance,'" Case 620-068 (Boston: Harvard Business School, April 2020, revised November 2020), 10.

24. Sara Lebow, "Google, Facebook, and Amazon to Account for 64% of U.S. Digital Ad Spending This Year," Insider Intelligence, November 3, 2021, https://www.emarketer.com/content/google-facebook-amazon -account-over-70-of-us-digital-ad-spending.

25. Annie Palmer, "Amazon Is Piling Ads into Search Results and Top Consumer Brands Are Paying Up for Prominent Placement," CNBC News, September 19, 2021, https://www.cnbc.com/2021/09/19/amazon-piles-ads -into-search-results-as-big-brands-pay-for-placement.html.

26. Louise Matsakis, "Small Businesses Say They Are Hurt by Rising Costs to Advertise on Amazon," NBC News, February 20, 2022, https://www.nbcnews.com/tech/small-businesses-say-are-hurt-rising -costs-advertise-amazon-rcna16685.

27. Bernard Marr, "The Amazing Ways Coca-Cola Uses Artificial Intelligence and Big Data to Drive Success," *Forbes*, September 18, 2017, https://www.forbes.com/sites/bernardmarr/2017/09/18/the-amazing-ways -coca-cola-uses-artificial-intelligence-ai-and-big-data-to-drive-success.

28. Marr, "The Amazing Ways Coca-Cola Uses Artificial Intelligence."

29. Tanya Dua, "How Coca-Cola Targeted Ads Based on People's Facebook, Instagram Photos," Digiday, May 16, 2017, https://digiday.com /marketing/coca-cola-targeted-ads-based-facebook-instagram-photos/.

30. Marr, "The Amazing Ways Coca-Cola Uses Artificial Intelligence."

31. Deborah Mary Sophia, "Kroger's Digital Push to Drive 2022 Sales, Profit Higher; Shares Jump," Reuters, March 3, 2022, https://www.reuters .com/business/retail-consumer/kroger-beats-quarterly-sales-estimates -pandemic-fueled-grocery-demand-sustains-2022-03-03/.

32. Alexander Coolidge, "Kroger's 84.51° Acquires Firm," *Cincinnati Enquirer*, August 1, 2016, https://www.cincinnati.com/story/money /2016/08/01/krogers-8451-acquires-firm/87912230/; Kannan Ramaswamy, William Youngdahl, and Kelly Molera, "The Digital Transformation of Kroger: Remaking the Grocery Business," Case TB0-636 (Boston: Harvard Business School, 2021).

33. Coolidge, "Kroger's 84.51° Acquires Firm."

34. Clayton M. Christensen, Rory McDonald, Laura Day, and Shaye Roseman, "Integrating Around the Job to Be Done," Case 611-004 (Boston: Harvard Business School, 2010).

35. "DBS Honoured as 'World's Best Bank' for Fourth Straight Year," DBS Bank, https://www.dbs.com/newsroom/DBS_honoured_as _Worlds_Best_Bank_for_fourth_straight_year_sg.

36. Vinika D. Rao, "How DBS Became the 'World's Best Bank,'" *INSEAD Knowledge*, November 15, 2021, https://knowledge.insead.edu /node/17671/pdf.

37. Robin Speculand, *World's Best Bank: A Strategic Guide to Digital Transformation* (Orlando, FL: Bridges Business Consultancy, 2021), 99–100.

38. "DBS Launches Southeast Asia's Largest Bank-Led Property Marketplace," DBS Bank, https://www.dbs.com/newsroom/DBS_launches _Southeast_Asias_largest_bank_led_property_marketplace.

39. Janice Lim, "DBS Launches Property Marketplace, Its Third Online Consumer Portal," Today Online, July 24, 2018, https://www.todayonline .com/singapore/dbs-launches-property-marketplace-its-third-online -consumer-portal.

40. Lim, "DBS Launches Property Marketplace."

41. "DBS to Roll Out 'Live More, Bank Less' Rebrand as Digital Transformation Takes Hold," Finextra, https://www.finextra.com /pressarticle/73937/dbs-to-roll-out-live-more-bank-less-rebrand-as -digital-transformation-takes-hold.

42. "DBS Full-Year Net Profit Rises 20% to Record SGD 8.19 Billion," DBS press release [no date], https://www.dbs.com/newsroom/DBS_full _year_net_profit_rises_20pct_to_record_SGD_8_19_billion.

43. Noah Barksy, "Nike's Earnings Calls Provide a Winning Digital Transformation Playbook," *Forbes*, July 27, 2021, https://www.forbes .com/sites/noahbarsky/2021/07/27/nike-earnings-calls-provide -winning-digital-transformation-playbook; "Helping to Make Sport a Daily Habit," Nike News press release, https://news.nike.com/news/nike-digital -health-activity-resources.

44. "Nike Studios," Nike, https://www.nike.com/nikestudios.

45. Cara Salpini, "Nike Plans 'Network' of Boutique Fitness Studios," *Retail Dive*, August 2, 2023, https://www.retaildive.com/news/nike -boutique-fitness-studios-training-running/689705/.

46. Armondo Tinoco and Dade Hayes, "Nike Training Club Launches on Netflix," *Deadline*, December 30, 2022, https://deadline.com/2022/12 /nike-training-club-launches-netflix-1235209159/.

47. Marco Iansiti, "The Value of Data and Its Impact on Competition," working paper 22-002, Harvard Business School, Boston, 2021, https://www.hbs.edu/ris/Publication%20Files/22-002submitted_835f63fd -d137-494d-bf37-6ba5695c5bd3.pdf, 3, 5.

48. Iansiti, "The Value of Data," 5.

49. Andrei Prakharevich, "Long-Tail Keywords: Find and Use Them for SEO," SEO PowerSuite, February 23, 2021, https://www.link-assistant.com /news/long-tail-keywords.html.

Chapter 3

1. "The App Store Turns 10," Apple press release, July 5, 2018, https://www.apple.com/newsroom/2018/07/app-store-turns-10/.

2. Christina Bonnington, "5 Years On, the App Store Has Forever Changed the Face of Software," *Wired*, July 10, 2013, https://www.wired .com/2013/07/five-years-of-the-app-store/.

3. "Apple Announces App Store Small Business Program," Apple press release, November 18, 2020, https://www.apple.com/newsroom/2020/11 /apple-announces-app-store-small-business-program/.

4. Feng Zhu and Nathan Furr, "Products to Platforms: Making the Leap," *Harvard Business Review*, April 2016.

5. Andy Chalk, "Valve Paid $20,000 to Hacker Who Discovered Critical Steam Security Flaw," *PC Gamer*, November 13, 2018, https://www .pcgamer.com/valve-paid-dollar20000-to-hacker-who-discovered-critical -steam-security-flaw/; Feng Zhu and Nathan Furr, "Products to Platforms: Making the Leap," *Harvard Business Review* (April 2016): 73–78.

6. Jeffrey Rousseau, "Valve: During 2021 Steam Saw 2.6m First-Time Buyers Each Month," GamesIndustry.biz, March 9, 2022, https://www .gamesindustry.biz/articles/2022-03-09-valve-during-2021-steam-saw -2-6m-first-time-buyers-each-month.

7. "Bloomberg Billionaires Index," *Bloomberg*, https://www .bloomberg.com/billionaires/profiles/gabe-newell/#xj4y7vzkg.

8. Sarah Perez, "Zelle, the Real-Time Venmo Competitor Backed by over 30 U.S. Banks, Arrives This Month," TechCrunch, June 12, 2017, https://techcrunch.com/2017/06/12/zelle-the-real-time-venmo -competitor-backed-by-over-30-u-s-banks-arrives-this-month/.

9. Perez, "Zelle, the Real-Time Venmo Competitor."

10. Theresa Stevens, "Zelle vs. Venmo: Which to Use and When," *Forbes Advisor*, October 11, 2022, https://www.forbes.com/advisor/money-transfer /Zelle-vs-venmo/.

11. "Zelle Soars with $80 Billion Transaction Volume, Up 28% from Prior Year," Zelle press release, March 4, 2024, https://www.zellepay.com /press-releases/zelle-soars-806-billion-transaction-volume-28-prior-year.

12. Nunez, "Zelle Records Explosive Q2 Growth."

13. "Zelle Soars with $80 Billion Transaction Volume."

14. Chiara Farronato, Stefano Denicolai, and Sarah Mehta, "Telepass: From Tolling to Mobility Platform," Case 622-011 (Boston: Harvard Business School, September 2021, revised December 2021).

15. Farronato et al., "Telepass."

16. "Amid the Shift from Print to Digital Learning, Simon Allen, CEO of McGraw Hill, Sees Plentiful Opportunities Ahead," CEO North America, https://ceo-na.com/executive-interviews/a-digital-transformation-in -education.

17. Bloomberg, "McGraw-Hill Education CEO on Digital Transformation," video, https://www.bloomberg.com/news/videos/2017 -09-27/mcgraw-hill-education-ceo-on-digital-transformation-video ?sref=gzAxNo0s.

18. "McGraw Hill and TutorMe Partner to Offer Free On-Demand Tutoring to Millions of College Students," McGraw-Hill press release, July 9, 2020, https://www.mheducation.com/news-media/press-releases /mcgraw-hill-virtual-tutoring-session.html.

19. Andy Wu, Feng Zhu, Pippa Tubman Armerding, and Wale Lawal, "EbonyLife Media (B)," Harvard Business School Supplement 722-378 (Boston: Harvard Business Review, November 2021, revised December 2021).

20. Farronato et al., "Telepass."

21. Farronato et al., "Telepass."

22. Melissa Repko, "Asian Grocery Start-up Weee Draws Shoppers with Tradition, Tech and a Dash of Hollywood," CNBC, May 20, 2022, https://www.cnbc.com/2022/05/20/weee-taps-crazy-rich-asians-director -jon-m-chu-in-push-for-grocery-growth.html.

23. "Instacart Launches 'Instacart Platform' with New Advertising, Fulfillment and Insights Solutions for Retailers," Instacart press release, March 23, 2022, https://www.instacart.com/company/pressreleases /instacart-launches-instacart-platform-with-new-advertising-fulfillment -and-insights-solutions-for-retailers/.

Chapter 4

1. Salman Haqqi, "The Most Popular Fashion Brands around the World," *Money*, November 10, 2021, https://www.money.co.uk/credit -cards/most-popular-fashion-brands.

2. Tao Lue Yang Qian, "This Chinese Company, Which You Have Never Heard Of, Has Become the Top of the Global Fashion Industry," PKU Speech, July 20, 2021, http://www.hsmrt.com/article/10690.

3. Kristoffer Tigue, "Black Friday's 'Enormous Environmental Impact' Sparks a Green Backlash," *Inside Climate News*, November 29, 2022, https://insideclimatenews.org/news/29112022/black-fridays-enormous -environmental-impact-sparks-a-green-backlash/; Sangeeta Singh -Kurtz, "Shein Is Even Worse Than You Thought," *The Cut*, October 17, 2022, https://www.thecut.com/2022/10/shein-is-treating-workers-even -worse-than-you-thought.html; Anna Papadopoulos, "The World's Most

Valuable Unicorns, 2023," *CEO World*, April 21, 2023, https://ceoworld
.biz/2023/04/21/the-worlds-most-valuable-unicorns-2023/.

4. Louise Matsakis, Meaghan Tobin, and Wency Chen, "How
Shein Beat Amazon at Its Own Game—And Reinvented Fast Fashion,"
Guardian, December 21, 2021, https://www.theguardian.com/fashion
/2021/dec/21/how-shein-beat-amazon-at-its-own-game-and-reinvented
-fast-fashion.

5. Huang Shan, "The Secret of Shein Selling Tens of Millions of Dollars
Every Day around the World Is Hidden in More Than 300 Factories in
Guangzhou," *Interface News*, August 17, 2021, https://fashion.sina.com
.cn/s/in/2021-08-17/1004/doc-ikqciyzm1910861.shtml.

6. Bruce Einhorn, "Shein's $100 Billion Value Would Top H&M and
Zara Combined," *Bloomberg*, April 4, 2022, https://www.bloomberg.com
/news/articles/2022-04-04/shein-s-100-billion-valuation-would-top-h-m
-and-zara-combined; "Shein: Riding the Wave of Cross-Border Ecommerce
[in Chinese]," *CITIC Securities*, July 11, 2021.

7. Lora Jones, "Shein: The Secretive Chinese Brand Dressing Gen Z,"
BBC, November 9, 2021, https://www.bbc.com/news/business-59163278.

8. Jones, "Shein."

9. Morgan Stanley Research, "Shein: Disrupting Fast Fashion,"
October 19, 2021, 19, retrieved from Bloomberg.

10. Olivia Rockeman, "Shein's US Expansion Adds Pressure for Its Fast-
Fashion Competitors," *Bloomberg*, November 1, 2022, https://www
.bloomberg.com/news/articles/2022-11-01/shein-delivery-times-to
-decrease-with-warehouses-opening.

11. Olivia Rockeman and Devon Pendleton, "Billionaire Claure
Expands Shein's Fast-Fashion Empire in Brazil," *Bloomberg*, May 23, 2023,
https://www.bloomberg.com/news/articles/2023-05-23/shein-adds-100
-factories-in-brazil-as-billionaire-grows-fast-fashion-empire.

12. Michelle Russell, "Shein Partners China Airline to Strengthen
Logistics Capacity," August 2, 2022, *JustStyle*, https://www.just-style.com
/news/shein-partners-china-airline-to-strengthen-logistics-capacity/.

13. Daomin Liu and Xinyao Liao, "Crossborder E-Commerce Report:
Shein [in Chinese]," *Sinolink Securities*, December 20, 2021, https://pdf
.dfcfw.com/pdf/H3_AP202112211535831166_1.pdf?1640081621000.pdf, 17.

14. Matsakis et al., "How Shein Beat Amazon."

15. Keith Zhai, "Fast-Fashion Giant Shein Explores Becoming Online
Marketplace," *Wall Street Journal*, December 12, 2022, https://www
.wsj.com/articles/fast-fashion-giant-Shein-explores-becoming-online
-marketplace-11670827480.

16. Qian, "This Chinese Company You Have Never Heard Of."

17. Packy McCormick, "Shein: The TikTok of Ecommerce," *Not Boring*,
May 17, 2021, https://www.notboring.co/p/shein-the-tiktok-of
-ecommerce.

18. "Shein's Market Strategy: How the Chinese Fashion Brand Is Conquering the West," Daxue Consulting, July 6, 2022, https://daxueconsulting.com/Shein-market-strategy/.

19. Nick Stratt, "How Anker Is Beating Apple and Samsung at Their Own Game," *The Verge*, May 22, 2017, https://www.theverge.com/2017/5/22/15673712/anker-battery-charger-amazon-empire-steven-yang-interview.

20. Juozas Kaziukėnas, "Amazon-Native Brand Anker Goes Public," Marketplace Pulse, August 25, 2020, https://www.marketplacepulse.com/articles/amazon-native-brand-anker-goes-public; Anker Innovations Technology Co. Market Cap data retrieved from Yahoo! Finance, https://finance.yahoo.com/quote/300866.SZ?p=300866.SZ.

21. "Day One: Stories of Entrepreneurship | Steven Yang, Anker Technology," YouTube, February 4, 2016, https://www.youtube.com/watch?v=IEhL4B5qWEE.

22. Jiangyong Lu, Shanfeng Zhang, and Xuanli Xie, "Innovation-Driven International Entrepreneurship: The Case of Anker Innovations," Peking University Case Study, 2023.

23. "About Us," Anker, https://en.anker-in.com/about/; "iF Design Award," 2022, https://ifdesign.com/en/brands-creatives/company/anker-gebrschoeller-gmbh-co-kg/2130.

24. "2021 Annual Report of Anker Innovations," Anker, 2021, http://file.finance.sina.com.cn/211.154.219.97:9494/MRGG/CNSESZ_STOCK/2022/2022-4/2022-04-12/7976955.PDF, 18; Juozas Kaziukėnas, "Anker's Sales Hit $1B on Amazon," Marketplace Pulse, June 7, 2022, https://www.marketplacepulse.com/articles/amazon-native-brand-anker-reaches-1-billion-sales.

25. "Apple's Only Cooperating Mainland Charging Brand Anker Is a Good Partner for iPhone 13," iNEWS, June 15, 2023, https://inf.news/en/digital/f4f6c708b6ef538eb12886b0423e953b.html.

26. Portions of this section are adapted from Feng Zhu, Anthony K. Woo, and Nancy Huai Dai, "Ping An: Pioneering the New Model of 'Technology-Driven Finance,'" Case 620-068 (Boston: Harvard Business School, April 2020, revised November 2020).

27. Ping An, 2017 Annual Report, http://www.pingan.com/app_upload/images/info/upload/fefe8a8e-fd10-4814-b7b2-aaecf814ff6d.pdf.

28. Feng Zhu, Anthony K. Woo, and Nancy Huai Dai, "Ping An: Pioneering the New Model of 'Technology-Driven Finance,'" Case 620-068 (Boston: Harvard Business School, April 2020, revised November 2020).

29. "Autohome at a Glance," Autohome, https://ir.autohome.com.cn/about-us.

30. "Auto Services Ecosystem," Ping An, https://group-test.pingan.com/about_us/our_businesses/auto-services-ecosystem.html.

31. Julie Zhu and Kane Wu, "Ping An Seeks to Sell $2.1 Bln Stake in Autohome—Sources," Reuters, November 12, 2021, https://www.reuters .com/business/exclusive-ping-an-seeks-sell-21-bln-stake-autohome-sources -2021-11-12/.

32. Zhu and Wu, "Ping An Seeks to Sell."

33. Zhu et al., "Ping An: Pioneering the New Model."

34. "Audited Annual Results for the Year Ended 31 December 2021," Ping An Healthcare and Technology Company, March 15, 2022, https:// portalvhds1fxb0jchzgjph.blob.core.windows.net/press-releases -attachments/1393371/HKEX-EPS_20220315_10154836_0.PDF.

35. Tom Marling, "Ping An Drops Property Listings Business," AIM Group, February 25, 2019, https://aimgroup.com/2019/02/25/ping-an -drops-property-listings-business-2/.

36. Xiang Guoliang, "The Death of Ping An Haofang [in Chinese]," Leju Finance, February 14, 2019, https://www.lejucaijing.com/news -6501739648329760168.html.

37. "Corporate Overview," Beike, https://investors.ke.com/.

38. "DBS Launches Southeast Asia's Largest Bank-Led Property Marketplace," DBS press release, July 24, 2018, https://www.dbs.com /newsroom/DBS_launches_Southeast_Asias_largest_bank_led _property_marketplace.

39. "Adidas Running: Run Tracker," Google Play, https://play.google .com/store/apps/details?id=com.runtastic.android&hl=en_US&gl=US.

40. "Xiaomi—the World's Largest IoT Platform," press release, April 6, 2018, miot-global.com website, https://miot-global .com/news-and-actions/xiaomi-the-worlds-largest-iot-platform/.

41. Xiaomi IPO Prospectus, 2018, http://cnbj1.fds.api.xiaomi.com /company/financial/en-US/IPO.pdf.

42. Ella Cao, "Xiaomi's CEO Announces Upgraded Core Strategy for Next Decade: Mobile x AIoT," *Pandaily*, August 17, 2020, https://pandaily .com/xiaomis-ceo-announces-upgraded-core-strategy-for-next-decade -mobile-x-aiot/.

43. "Xiaomi Again Advances on the Fortune Global 500 List," press release, August 3, 2022, http://en.ccceu.eu/2022-08/12/c_2366.htm.

44. See Ron Adner, *The Wide Lens: A New Strategy for Innovation* (New York: Portfolio, 2013), chapter 2.

45. Shane Greenstein, Feng Zhu, and Kerry Herman, "Korea Telecom: Building a GIGAtopia (A)," Case 617-014 (Boston: Harvard Business School, April 2017, revised January 2020).

46. Greenstein et al., "Korea Telecom," 9–10.

47. Alan Weissberger, "South Korea Has 30 Million 5G Users, but Did Not Meet Expectations; KT and SKT AI Initiatives," IEEE ComSoc Technology Blog, May 9, 2023, https://techblog.comsoc.org/2023/05/09/south-korea-has -30-million-5g-users-but-did-not-meet-expectations-kts-5g-ai/.

48. "Ping An Launches Home-Based Elderlycare Service to Pursue 'Finance + Elderlycare' Business Strategy," press release, March 11, 2022, https://www.prnewswire.com/news-releases/ping-an-launches-home -based-elderlycare-service-to-pursue-finance--elderlycare-business -strategy-301500771.html.

49. "WeChat Speaks: Chinese Say 'No' to Nursing Homes," Collective Responsibility, July 7, 2016, https://www.coresponsibility.com/wechat -chinese-say-no-nursing-homes/.

Chapter 5

1. Feng Zhu and Angela Acocella, "X Fire Paintball & Airsoft: Is Amazon a Friend or Foe? (A)," Case 617-046 (Boston: Harvard Business School, January 2017, revised August 2019).

2. Giulia Morpurgo and Antonia Vanuzzo, "Domino's Pizza Quits Italy after Locals Shun American Pies," *Bloomberg*, August 9, 2022, https://www.bloomberg.com/news/articles/2022-08-09/domino-s-pizza -leaves-italy-as-traditional-margherita-wins?sref=gzAxNo0s.

3. Feng Zhu, Krishna G. Palepu, Bonnie Yining Cao, and Dawn H. Lau, "Pinduoduo," Case 620-040 (Boston: Harvard Business School, 2019).

4. Binnie Wong and Chen Wang, "Pinduoduo Initiation Report," HSBC Global Research, January 31, 2019, retrieved from Bloomberg.

5. "Amazon (AMZN) Q2 2022 Earnings Call Transcript," Motley Fool, July 28, 2022, https://www.fool.com/earnings/call-transcripts/2022/07/28 /amazon-amzn-q2-2022-earnings-call-transcript/.

6. Dana Mattioli, "Amazon Scooped Up Data from Its Own Sellers to Launch Competing Products," *Wall Street Journal*, April 23, 2020, https://www.wsj.com/articles/amazon-scooped-up-data-from-its-own -sellers-to-launch-competing-products-11587650015.

7. Dana Mattioli, "Amazon Changed Search Algorithm in Ways That Boost Its Own Products," *Wall Street Journal*, September 16, 2019, https://www.wsj.com/articles/amazon-changed-search-algorithm-in-ways -that-boost-its-own-products-11568645345.

8. Kanishka Singh, "Uber Eats, DoorDash, Grubhub Sue New York City over Legislation on Commission Caps," Reuters, September 10, 2021, https://www.reuters.com/technology/grubhub-doordash-uber-eats-sue -new-york-city-over-fee-caps-wsj-2021-09-10/.

9. Tommy Pan Fang, "Managing Platform Value through Business Model Governance," HKU Business School online seminar, September 24, 2021; D. Daniel Sokol and Feng Zhu, "Harming Competition and Consumers under the Guise of Protecting Privacy: An Analysis of Apple's iOS 14 Policy Updates," *Cornell Law Review Online* 101, no. 3 (June 1, 2021), https://papers.ssrn.com/sol3/papers.cfm?abstract_id=3852744.

10. Preetika Rana, "DoorDash Sues New York City over Sharing Data with Restaurants," *Wall Street Journal*, September 15, 2021, https://www.wsj.com/articles/doordash-sues-new-york-city-over-sharing-data-with-restaurants-11631711461.

11. Aneurin Canham-Clyne, "Grubhub Sues NYC over Restaurant Data-Sharing Law," *Restaurant Dive*, December 13, 2021, https://www.restaurantdive.com/news/grubhub-sues-nyc-over-restaurant-data-sharing-law/611389/.

12. Huang Yichang, "Conflict between SF and Cainiao Hits Delivery Service," CGTN, June 2, 2017, https://news.cgtn.com/news/3d67444d7a45444e/share_p.html.

13. "Chinese Billionaires Clash over Alibaba's Parcel Deliveries," *Bloomberg*, June 1, 2017, https://www.bloomberg.com/news/articles/2017-06-02/chinese-billionaires-clash-over-alibaba-s-parcel-deliveries.

14. Ben Zimmerman, "Why Nike Cut Ties with Amazon and What It Means for Other Retailers," *Forbes*, January 22, 2020, https://www.forbes.com/sites/forbesbusinesscouncil/2020/01/22/why-nike-cut-ties-with-amazon-and-what-it-means-for-other-retailers/?sh=7cbb3b8e64ff.

15. Phil Wahba, "Gap Would Consider Using Amazon to Help Fix Its Sales Problem," *Fortune*, May 18, 2016, https://fortune.com/2016/05/18/gap-amazon-sales/.

16. Tram News, "SAIC Refuses to Cooperate with Huawei on Autonomous Driving and Wants to Keep the Soul in Its Own Hands," *OF Week*, July 5, 2017, https://nev.ofweek.com/2021-07/ART-77015-8500-30507885.html.

17. Chris Ziegler, "GM Aims to Speed Up Self-Driving Car Development by Buying Cruise Automation," *The Verge*, May 11, 2016, https://www.theverge.com/2016/3/11/11195808/gm-cruise-automation-self-driving-acquisition.

18. Joseph White, "GM Plans to Phase Out Apple CarPlay in EVs, with Google's Help," Reuters, March 31, 2023, https://www.reuters.com/technology/gm-plans-phase-out-apple-carplay-evs-with-googles-help-2023-03-31/.

19. Jacob Kastrenakes, "Disney to End Netflix Deal and Launch Its Own Streaming Service," *The Verge*, August 8, 2017, https://www.theverge.com/2017/8/8/16115254/disney-launching-streaming-service-ending-netflix-deal.

20. Portions of this section are adapted from Feng Zhu, Yulin Fang, Bonnie Yining Cao, and Duan Yang, "Huazhu: A Chinese Hotel Giant's Journey of Digital Transformation," Case 622-071 (Boston: Harvard Business School, February 2022, revised January 2023).

21. "Hotel Industry: New Period Outlook, the Leading Highlight Is High Certainty and High Flexibility [in Chinese]," Northeast Securities Co., Ltd., November 20, 2020, 28, https://bigdata-s3.wmcloud.com/mailreport/mailContent/1c51c814089fb296fa2237d3a22cbc4c.pdf.

22. "Huazhu Group 2021 Full Year Earnings Call Presentation," Huazhu Group, March 24, 2022, https://ir.hworld.com/static-files/9b894ed8-4585 -4c4c-936f-9fbd4d9ee8f8, 14–15.

23. Zhu et al., "Huazhu: A Chinese Hotel Giant's Journey of Digital Transformation."

24. Portions of this section are adapted from Feng Zhu and Angela Acocella, "X Fire Paintball & Airsoft: Is Amazon a Friend or Foe? (B)," Case 617-047 (Boston: Harvard Business School, January 2017, revised August 2019).

25. Zhu and Acocella, "X Fire Paintball & Airsoft: Is Amazon a Friend or Foe? (B)."

26. Samuel Axon, "Amazon and Best Buy Team Up to Sell TVs, but It's a Risky Move for Best Buy," April 18, 2018, https://arstechnica.com /gadgets/2018/04/longtime-rivals-amazon-and-best-buy-announce -a-major-tv-partnership/.

27. "Domino's Introduces a New Way to Order Using Uber Eats Marketplace," Domino's press release, July 12, 2023, https://ir.dominos .com/news-releases/news-release-details/dominosr-introduces-new-way -order-using-uber-eats-marketplace.

28. Margaret Harding McGill, "Scoop: Anti-Apple Coalition Taps New Leader," Axios, February 10, 2021, https://www.axios.com/2021 /02/10/coalition-for-app-fairness-meghan-dimuzio-exec-director.

29. Margaret Sullivan, "These Local Newspapers Say Facebook and Google Are Killing Them. Now They're Fighting Back," Washington Post, February 2, 2021, https://www.washingtonpost.com/lifestyle/media/west -virginia-google-facebook-newspaper-lawsuit/2021/02/03/797631dc-657d -11eb-8468-21bc48f07fe5_story.html.

30. Harro Ten Wolde and Eric Auchard, "Germany's Top Publisher Bows to Google in News Licensing Row," Reuters, November 5, 2014, https://www.reuters.com/article/us-google-axel-sprngr/germanys-top -publisher-bows-to-google-in-news-licensing-row-idUSKBN0IP1YT20141105.

31. Wolde and Auchard, "Germany's Top Publisher Bows to Google."

32. Treasury Laws Amendment (News Media and Digital Platforms Mandatory Bargaining Code) Bill 2021, The Parliament of the Commonwealth of Australia, House of Representatives, https://www.accc .gov.au/system/files/Final%20legislation%20as%20passed%20by%20 both%20houses.pdf.

33. Morgan Meaker, "Australia's Standoff against Google and Facebook Worked—Sort Of," Wired, February 25, 2022, https://www.wired.com /story/australia-media-code-facebook-google/.

34. Johannes Munter, "Australia's News Media Bargaining Code Is a Major Success That the U.S. Can Emulate," News Media Alliance, August 25, 2022, https://www.newsmediaalliance.org/australias-news -media-bargaining-code-is-a-major-success-that-the-u-s-can-emulate/.

35. Matthew Ingram, "Canada Imitates Australia's News-Bargaining Law, but to What End?" *Columbia Journalism Review,* March 16, 2023, https://www.cjr.org/the_media_today/canada_australia_platforms _news_law.php; Scott Roxborough, "Google News Reopens in Spain after 8-Year Break, Drops Ad Revenue Fight in France," *The Hollywood Reporter,* June 22, 2022, https://www.hollywoodreporter.com/business /digital/google-news-reopens-in-spain-reaches-agreement-with-french -publishers-1235169822/.

36. Sarah Fischer and Kristal Dixon, "Scoop: Over 200 Papers Quietly Sue Big Tech," *Axios,* December 7, 2021, https://www.axios.com/2021 /12/07/1-local-newspapers-lawsuits-facebook-google.

37. Journalism Competition and Preservation Act of 2022, S. 673, 117th Congress, 2nd Edition, Calendar No. 569.

38. Amy Klobuchar, "Klobuchar, Kennedy, Cicilline, Buck, Durbin, Nadler Release Updated Bipartisan Journalism Bill," Amy Klobuchar Senate, August 22, 2022, https://www.klobuchar.senate.gov/public/index .cfm/2022/8/klobuchar-kennedy-cicilline-buck-durbin-nadler-release -updated-bipartisan-journalism-bill.

39. Dean Ridings, "Antitrust Bill Necessary to Protect Local News from Google and Facebook," *Baltimore Sun,* July 20, 2022, https://www .baltimoresun.com/opinion/op-ed/bs-ed-op-0720-jcpa-20220720 -ceznmreg7vakfcmxljrwn5wyba-story.html.

Chapter 6

1. Laura He, "Top Chinese Education Company Laid Off 60,000 People after Beijing's Crackdown Last Year," CNN, January 10, 2022, https://www.cnn.com/2022/01/10/business/china-education-layoff-new -oriental-intl-hnk/index.html.

2. "FY21Q1 Fact Sheet," New Oriental, https://investor.neworiental .org/static-files/f72cfa57-d68c-4db1-b7f0-13446a7c8b54.

3. GETChina Insights, "Koolearn, Subsidiary of New Oriental, Made the First Online Education Stock in Hong Kong," Medium, April 1, 2019, https://edtechchina.medium.com/koolearn-subsidiary -of-new-oriental-made-the-first-online-education-stock-in-hong-kong -ba841890014a.

4. Shen Lu, "Targeted by Beijing, One Chinese Tutoring Company Reinvents Itself with Live Streams Selling Groceries," *Wall Street Journal,* July 13, 2022, https://www.wsj.com/articles/targeted-by-beijing-one -chinese-tutoring-company-reinvents-itself-with-live-streams-selling -groceries-11657704780.

5. David Goldman, "Blockbuster Is 'Bleeding to Death,'" CNNMoney, March 24, 2010, https://money.cnn.com/2010/03/24/news/companies /blockbuster/?npt=NP1.

6. Tom Huddleston Jr., "Netflix Didn't Kill Blockbuster—How Netflix Almost Lost the Movie Rental Wars," CNBC, September 22, 2020, https://www.cnbc.com/2020/09/22/how-netflix-almost-lost-the-movie -rental-wars-to-blockbuster.html.

7. Anders Melin and Bryan Gruley, "Who's a Very Good Pandemic Business? Chewy Is. Oh, Yes, It Is," *Bloomberg*, November 18, 2020, https://www.bloomberg.com/news/features/2020-11-18/chewy-chwy-is -having-a-great-2020-as-pet-care-surges-in-the-pandemic?sref=gzAxNo0s.

8. Marissa Heflin, "Chewy Turns 10: How the Online Retailer Has Impacted the Pet Industry," *Pet Product News*, July 21, 2021, https://www .petproductnews.com/news/chewy-turns-10-how-the-online-retailer -has-impacted-the-pet-industry/article_c91c4f0a-ea43-11eb-9b30 -d3e8047c206e.html.

9. Miriam Gottfried, "How PetSmart Swallowed Chewy—and Proved the Doubters Wrong," *Wall Street Journal*, October 1, 2019, https://www .wsj.com/articles/how-petsmart-swallowed-chewyand-proved-the -doubters-wrong-11569858310.

10. Gottfried, "How PetSmart Swallowed Chewy."

11. Emma Bedford, "Leading Pet Specialty Chains in North America as of March 2022, Based on Number of Stores," Statista, May 20, 2022, https://www.statista.com/statistics/253896/leading-north-american-pet -specialty-chains-by-number-of-stores/.

12. Lindsey Grant, "What PetSmart's Split from Chewy Means to the Industry," *Pet Product News*, October 28, 2020, https://www .petproductnews.com/news/what-petsmart-s-split-from-chewy-means-to -the-industry/article_eee7bb16-1942-11eb-b72c-dba94f5f466c.html.

13. "PetSmart Selling Its Stake in Chewy," *Pet Business*, October 26, 2020, https://www.petbusiness.com/industry-news/petsmart-selling-its -stake-in-chewy/article_c2e152ba-17b3-11eb-bec7-6b0c543a7542.html.

14. "Marriott International Marks Another Year of Strong Acceleration in Signings," Marriott press release, January 23, 2023, https://news .marriott.com/news/2023/01/23/marriott-international-marks-another -year-of-strong-acceleration-in-signings.

15. Sean O'Neill, "Marriott Bonvoy Adds 200 Millionth Member as Hotel Race Heats Up," February 26, 2024, https://skift.com/2024/02/26 /marriott-bonvoy-adds-200-millionth-member-as-hotel-loyalty-race -heats-up.

16. Joel Thomas, "2022 Airbnb Statistics: Usage, Demographics, and Revenue Growth," Stratos Jet Charters, January 4, 2022, https://www .stratosjets.com/blog/airbnb-statistics/.

17. David Silverberg, "Airbnb Turns to AI to Help Prevent House Parties," BBC News, October 2023, https://www.bbc.com/news/business -67156176.

18. "Proven Ways to Attract Business Travellers to Your Airbnb," Expert Easy, July 20, 2022, https://www.experteasy.com.au/blog/proven-ways

-to-attract-business-travellers-to-your-airbnb/; "Get Your Airbnb 'Business Travel Ready,'" GuestReady, https://www.guestready.com/blog/airbnb -business-travel-ready/.

19. Marvin Scholz, "Marriott Now Offers Short Term Rentals Competing with Airbnb," May 10, 2022, https://www.traveloffpath.com /marriott-now-offers-short-term-rentals-competing-with-airbnb/; Cameron Sperance, "Marriott Launches Its 32nd—and Most Affordable—Brand," The Points Guy, https://thepointsguy.com/news/marriott-budget -extended-stay-travel/.

20. "Airbnb 2019 Business Update," Airbnb, https://news.airbnb.com /airbnb-2019-business-update/.

21. We thank our colleague Krishna Palepu for sharing his research insights on this topic.

22. Juro Osawa, "Alipay Wallet Hits 190 Million Active Users," *Wall Street Journal*, October 15, 2014, https://www.wsj.com/articles/alipay -wallet-hits-190-million-active-users-1413430508.

23. Krishna G. Palepu, Feng Zhu, Susie L. Ma, and Kerry Herman, "Ant Group (A)," Case 122-003 (Boston: Harvard Business School, October 2021, revised February 2023).

24. Ant Group, H Share IPO, https://www1.hkexnews.hk/listedco /listconews/sehk/2020/1026/2020102600165.pdf, 196; Jing Yang and Xie Yu, "Jack Ma's Ant Group Ramped Up Loans, Exposing Achille's Heel of China's Banking System," *Wall Street Journal*, December 6, 2020, https://www.wsj.com/articles/jack-mas-ant-group-ramped-up-loans-exposing-achilles-heel-of-chinas-banking-system -11607250603.

25. Yang and Yu, "Jack Ma's Ant Group."

26. Julie Zhu and Jane Xu, "China Ends Ant Group's Regulatory Ramp with Nearly $1 Billion Fine," *Reuters*, July 7, 2023, https://www.reuters .com/technology/china-end-ant-groups-regulatory-revamp-with-fine-least -11-bln-sources-2023-07-07/.

27. Willy Shih, "The Real Lessons from Kodak's Decline," *MIT Sloan Management Review*, May 20, 2016, https://sloanreview.mit.edu/article /the-real-lessons-from-kodaks-decline/.

28. Oliver Kmia, "Why Kodak Died and Fujifilm Thrived: A Tale of Two Film Companies," PetaPixel, December 14, 2022, https://petapixel .com/2018/10/19/why-kodak-died-and-fujifilm-thrived-a-tale-of-two -film-companies; Christopher Sirk, "Fujifilm vs. Kodak: Navigating Digital Innovation & Survival," January 30, 2024, https://crm.org /articles/fujifilm-found-a-way-to-innovate-and-survive-digital-why -didnt-kodak.

29. Alex Knapp, "How Garmin Mapped Out a New Direction with Fitness Wearables," *Forbes*, September 14, 2016, https://www.forbes.com /sites/alexknapp/2016/09/14/how-garmin-mapped-out-a-new-direction -with-fitness-wearables/.

30. Quy Huy, "How Nokia Bounced Back (with the Help of the Board)," INSEAD Knowledge, October 10, 2018, https://knowledge.insead.edu /strategy/how-nokia-bounced-back-help-board.

31. "Nokia: The Only Viable Non-Chinese 5G Play," Fade the Market, March 24, 2022, https://seekingalpha.com/article/4497500-nokia -viable-non-chinese-5g-play.

Conclusion

1. Alicia Kelso, "How Becoming 'A Tech Company That Sells Pizza' Delivered Huge for Domino's," *Forbes*, April 30, 2018, https://www.forbes .com/sites/aliciakelso/2018/04/30/delivery-digital-provide-dominos-with -game-changing-advantages/.

2. Kara Carlson, "As Farming Goes High-Tech, John Deere Opens Development Facility in Austin," *Austin American-Statesman*, February 25, 2022, https://www.statesman.com/story/business/2022/02/25/john-deere -opens-austin-office-focused-agriculture-technology/6914135001/.

3. Anna Baydakova, "DBS Bank CEO: We Have Twice as Many Engineers as Bankers," Yahoo! Finance, May 26, 2021, https://finance .yahoo.com/news/dbs-bank-ceo-twice-many-210000778.html.

ACKNOWLEDGMENTS

Our investigation into how traditional businesses can survive and thrive against the challenges posed by tech giants started long before the generative AI hype emerged. The advent of ChatGPT, Gemini, and other large language models has greatly augmented the capabilities of these tech giants. Consequently, it has become even more crucial for traditional businesses to evolve into smart rivals.

We owe a tremendous debt of gratitude to the many individuals who have imparted their wisdom and insights throughout this journey.

Feng extends his heartfelt gratitude to his adviser, Marco Iansiti, who has been a pillar of support in all aspects of his life since the commencement of his doctoral studies at Harvard. Marco not only enlightened Feng on conducting business research but also sharpened Feng's ability to communicate insights to a wider audience. Karim Lakhani also played a crucial role in shaping Feng's research on digital innovation. Feng is also grateful for the guidance received from Shane Greenstein, with whom he has worked on several projects about Wikipedia, highlighting how digital innovation can disrupt established products, such as *Encyclopaedia Britannica*. Feng also cherishes the intellectual partnership with Amy Bernstein, who has consistently refined his ideas for greater clarity, impact, and managerial relevance.

Bonnie would like to thank her mentors from her journalism era—Linus Chua, Joe McDonald, Scott Tong, and Rob Urban. Their guidance and rigorous training not only shaped her career path but continue to serve as a source of inspiration to this day.

Our thanks and appreciation must also go to our colleagues at Harvard Business School (HBS) for their pivotal contributions to our understanding of the dynamics that drive the success of traditional businesses. We are deeply grateful to Prithwiraj Choudhury, Thomas Eisenmann, Ramon Casadesus-Masanell, Chiara Farronato, Kris Ferreira, Walter Frick, Sunil Gupta, Kerry Herman, Rory McDonald, Toni Moreno, Krishna Palepu, Stefan Thomke, Andy Wu, and the many others who have shared their knowledge and insights for the book. We also extend special thanks to our executive fellows, Philip Kuai and Jianwen Liao, for their invaluable collaboration.

Inspiration and support have also come from numerous scholars and friends beyond HBS. M. S. Krishnan, for example, inspired Feng to delve into the transformative journey of Domino's Pizza. Ruomeng Cui provided insightful feedback on initial drafts, while the steadfast support from Danny Sokol propelled Feng to see this project through to its completion.

We extend our thanks to our deans, Srikant Datar and Nitin Nohria, and many leaders at HBS, for fostering an environment that enabled our engagement with a myriad of traditional businesses through various educational programs. We also thank the numerous smart students and companies that we worked with. Collaborations with colleagues across HBS departments and our global research centers have further enriched our work. We are grateful to Pippa Armerding, Nancy Dai, Jaye Glenn, Pedro Levindo, Shu Lin, Fernanda Miguel, Tracy Pang, Adina Wong, and Sia Zhou.

The exceptional team at Harvard Business Review Press certainly deserves acknowledgment. Our editor, Kevin Evers, and production editor, Anne Starr, have been pillars of support, offering encouragement, invaluable advice, and empathy throughout this process. Our thanks also go to our copyeditor David Goehring, editorial coordinator Cheyenne Paterson, as well as our book jacket designer, Stephani Finks, for their outstanding contributions. The global sales and marketing team at HBR Press, including Jon Shipley, Julie Devoll, Felicia Sinusas, and Lindsey Dietrich have been dedicated to the book's promotion, for which we are deeply thankful.

Our families have our immense appreciation for their unwavering support and patience during the many late evenings and long weekends spent writing, time that could have been spent with them.

Feng is especially thankful to his wife, Ping, whose love and encouragement have been the cornerstone of his determination throughout this journey. She has infused the process with a wealth of ideas, passion, comfort, and countless smiles. Feng also thanks his son, Evan, whose keen interest in his research has been a constant motivator. Evan's spontaneous hugs have been delightful and welcome distractions. Feng is also grateful to his parents, Juanyu and Zudi, for their love and enduring support.

Bonnie's deepest gratitude is to her parents CJ and Ming for their unconditional love and support. She'd like to express her profound appreciation to her husband Charles and their new-born baby Eureka, who have been by her side and inspiring many "Eureka moments" during the book-writing adventure.

Finally, we extend special thanks to you, our readers. We hope this book enriches your lives and contributes to your success.

FENG ZHU is the MBA Class of 1958 Professor of Business Administration at Harvard Business School, where he also codirects the Platform Lab within the Digital, Data, and Design Institute and cochairs the Harvard Business Analytics Program. Zhu is renowned for his expertise in platform strategy, digital innovation, and competitive strategy. His research and teaching have won him many international awards. He has provided strategic advice on competition and regulation issues to leading technology companies, including Alphabet, Meta, Microsoft, and Uber. Additionally, Zhu has offered his expertise to a variety of traditional businesses, assisting them in navigating the digital landscape.

BONNIE YINING CAO is an award-winning former journalist for Bloomberg News in Shanghai and New York City and for the Associated Press in Beijing. She has written extensively about companies in tech and traditional industries in the United States and in emerging markets. Cao helped establish Bloomberg's real estate and hotel beat in China. Her work has appeared in *Bloomberg Businessweek*, the *Washington Post*, *Business Insider*, the *South China Morning Post*, and other outlets. She also worked as a researcher and case writer at Harvard Business School's Asia-Pacific Research Center.

Cao is currently pursuing her PhD in business administration at Harvard Business School. She holds a master's in public administration from Harvard's John F. Kennedy School of Government.

To learn more, visit www.smartrivals.com.